Tallgrass Prairie Restoration in the Midwestern and Eastern United States

Harold W. Gardner

Tallgrass Prairie Restoration in the Midwestern and Eastern United States

A Hands-On Guide

Springer

Harold W. Gardner
440 Mount Zion Road
Carlisle, PA 17015
USA
hrldgrdnr4@aol.com

Please note that additional material for this book can be downloaded from
http://extras.springer.com

ISBN 978-1-4419-7426-6 e-ISBN 978-1-4419-7427-3
DOI 10.1007/978-1-4419-7427-3
Springer New York Dordrecht Heidelberg London

Printed on acid-free paper

Springer is part of Springer Science+Business Media (www.springer.com)

Preface

In the early 1970s Irene Cull, a botanist trained at the University of Wisconsin, introduced me to some of the few remaining prairies in the area of Peoria, Illinois. These incredible prairies were reminiscent of increasingly rare native plant communities found in remote areas from Hawaii to the East coast. Whenever native plant communities are encountered, one can intuitively sense the authenticity by their inherent beauty and highly organized ecology. Unfortunately, ecological fragmentation occurs in the more populated regions over the United States. These areas are continually invaded with a myriad of weedy Eurasian plant introductions, even on preserved public land.

Irene Cull persuaded the management of the USDA lab in Peoria to landscape a portion of the laboratory property as native prairie. This project taught many lessons on prairie restoration, ranging in degrees of difficulty. Then, in the mid-1970s I extended Irene's tutelage to the restoration of degraded prairies in Jubilee State Park with the help of Sierra Club. About the same time row crops were started for seed production, both in the park and on private land. The row-crop project was the tool for eventually restoring 40+ acres of high-quality diverse prairie in the park and on private land. By the early 1980s a team of about 20 volunteers called "The Prairie Dawgs" was organized to focus on prairie restoration in the park. Controlled burn equipment and brush-cutting tools were purchased from proceeds of a prairie conference and grants from The Nature Conservancy, Sierra Club, and Illinois Department of Conservation. A regular burn schedule immensely improved the condition of about 250 acres of degraded prairies. As an added bonus, we began to realize that the burned forest areas showed a decrease in brush. Also, there was a loss of the lowest portion of the canopy, giving the feeling of an open or managed forest.

By the early 1990s a half-mile of Burlington Northern Railroad right-of-way near Brimfield, Illinois, became available for purchase. This remnant prairie had many excellent stands of prairie, but was degraded in the sense that it lacked diversity and contained alien species, particularly woody brush/trees, wild parsnip, and white sweet clover. After my wife and I purchased the right-of-way, aggressive tree and brush cutting was ongoing for several years. Persistent weeding virtually eliminated the parsnip within about 5 years, but white sweet clover remains a problem in small areas. To increase diversity, locally gathered seeds were scattered by hand.

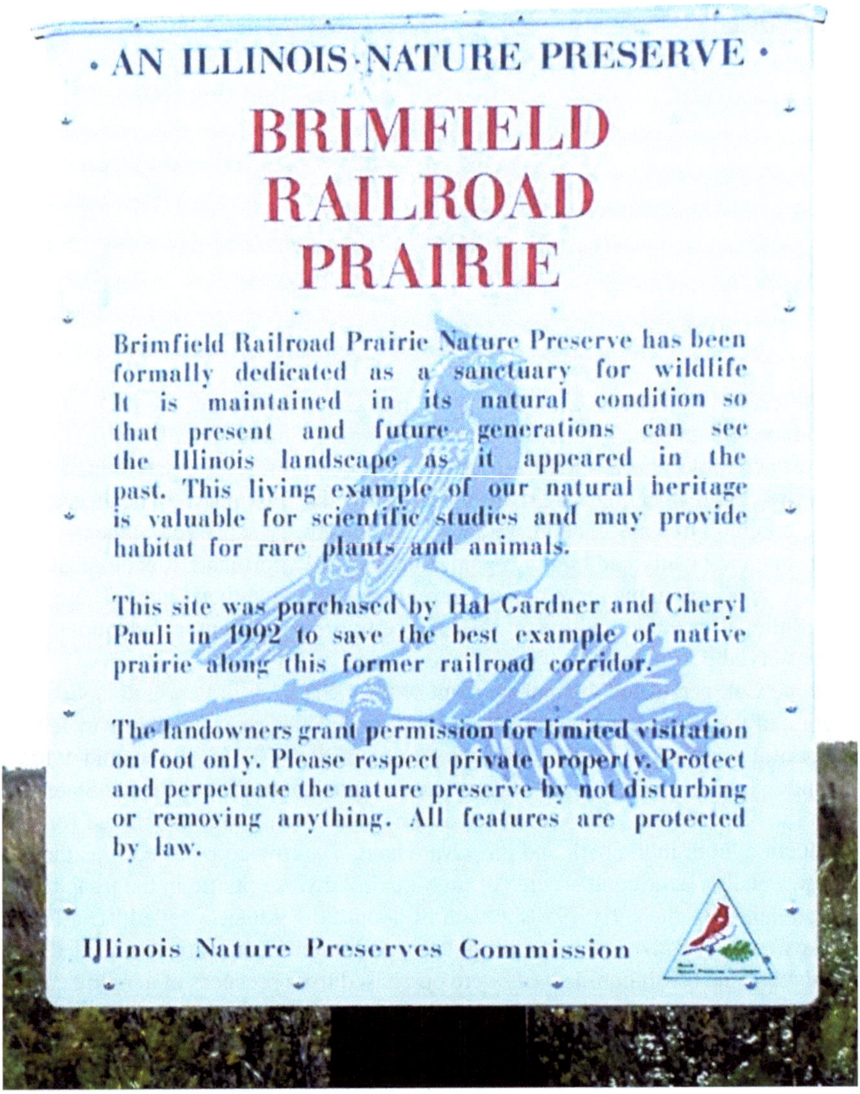

Brimfield Railroad Nature Preserve
Photo: Douglas Franks

It became evident that success in restoration by over-seeding an established perennial eco-system was more difficult than starting with weed-free tilled soil. Annual seeding coupled with time resulted in a dramatic increase in diversity to the extent that the property was designated an official Illinois Nature Preserve by the State of Illinois in 2002.

Several years ago, the USDA Natural Resources Conservation Service contracted with the author to restore 15+ acres as prairie in their Wildlife Habitat

Incentives Program (WHIP). The author continues to restore prairie areas on his farm in south central Pennsylvania where he maintains beds for seed production of about 150 species.

Also, the USDA NRCS sponsors a number of soil conservation initiatives (see Section 9). These USDA initiatives have resulted in a large nationwide increase in acreage put aside as grassland, both in the East and Midwest. On the downside, the USDA permits the use of Eurasian grasses as well.

Unfortunately, the public is under-educated about native plants. Even so-called garden experts consider the likes of Queen Anne's lace and dames rocket to be native wildflowers. Several years ago I was uprooting parsnip in the Rock Island Trail Nature Preserve. A hiker asked what I was doing. "Pulling weeds," I replied. He returned, "they are all weeds." The hiker seemed semi-satisfied only after I explained that I wasn't pulling "American weeds." The goal of this volume is to provide a comprehensive guide to all aspects of prairie restoration wherever conditions are favorable to restore these beautiful sun-loving native plants. Open fields in a large region from the East to the Midwest are very suitable for restoration.

Carlisle PA, USA Harold W. Gardner

Acknowledgments

Special thanks are accorded to Doug Franks, and Dr. Mark A. Berhow, both of Peoria, Illinois, for photos and scanning of prairie samples. Thanks are also due to Debra Grim for use of her photograph, and to Dr. Doug Doehlert of North Dakota State University for helpful comments. The author gratefully acknowledges the personnel of Jennings Prairie in Pennsylvania for their assistance. Thanks are also due to my former professor, Dr. Norman Deno, who taught me Mechanisms of Organic Chemistry at Penn State more than 40 years ago. Recently, Dr. Deno has once again served as a mentor by giving me access to his extensive seed germination studies. Thanks to my wife, Cheryl Pauli, and my son, Bryce Gardner, for tolerating my prairie obsession for many years.

Contents

Part I
Introduction

Chapter 1
Introduction to the Prairie

1.1 Why Prairie?

Prairie and savanna are among the most endangered of ecosystems. In Illinois, extant prairies account for only 0.01% of their original range [1]. Other Midwestern states are similarly deficient in prairie remnants. Because prairie soil in the Midwest is incredibly fertile, much of the former grasslands have disappeared under the plow. Interestingly, the root-bound prairie sods were intractable before the introduction of the John Deere steel moldboard plow. Several yoke of oxen were used to break the soil (Fig. 1.1). The snap of the breaking roots of lead plant (*Amorpha canescens*) earned lead plant the alternative name "prairie shoestring."

Even as early as the 1840s a great proportion of the tallgrass prairie had disappeared. The remaining prairies and savannas (widespread trees interspersed with prairie) are endangered because of the lack of fire. Thus, prairies are being transformed into areas of brush and trees. The former giant spreading oaks found on savannas are being degraded by younger forest trees, which shade the widespread limbs of full-sun trees. The decline of these ancient oaks is surprisingly rapid under these conditions.

What an amazing sight the tallgrass prairie must have been to the first people of European ancestry to view these undisturbed natural areas. Plant collection was one goal of the Lewis and Clark expedition. However, Thomas Nuttall, an apprentice printer from England, was much interested in botany. Nuttall endured incredible hardships collecting plants in the tallgrass area shortly after the Lewis and Clark expedition. He proceeded down the Ohio River, and subsequently traveled north, often by foot, through the heart of the tallgrass prairie. He discovered a large number of plants typical of native prairie absent or very rare in the East, such as *Amorpha canescens*, *Artemisa ludoviciana*, *Symphyotrichum* × *amethystinus*, *Symphyotrichum oblongifolium*, *Baptisia bracteata* var. (v.) *leucophaea*, *Arnoglossum plantagineum*, *Cirsium pumilum*, *Comandra umbellata*, *Coreopsis palmata*, *Delphinium carolinianum* v. *virescens*, *Echinacea pallida*, *Penstemon digitalis*, *Penstemon grandiflorus*, *Physostegia parviflora*, *Polytaenia nuttallii*, *Orbexilum onobrychis*, *Pycnanthemum verticillatum* v. *pilosum*, *Ruellia humilis*, *Mimosa microphylla*, and *Solidago speciosa*, among many others.

H.W. Gardner, *Tallgrass Prairie Restoration in the Midwestern and Eastern United States: A Hands-On Guide*, DOI 10.1007/978-1-4419-7427-3_1,

Fig. 1.1 Breaking the prairie sod at Bishop Hill, Illinois. "Breaking Prairie" by Olof Krans (courtesy Bishop Hill State Historic Site Museum)

Although the treeless prairie was a frightening unfamiliar terrain to early pioneers, there were some who appreciated their wondrous floral beauty. As reported by Betz [2], a visitor from New York in 1838 wrote:

> One who has never looked upon the Western prairie in the pride of its blushing bloom can hardly conceive the surpassing loveliness of its summer flora; and if the idea is not easy to conceive, still less is it so to convey. The autumn flowers in their richness I have not yet beheld; and in the early days of June, when I first stood upon the prairies, the beauteous sisterhood of spring were all in their graves; and the sweet springtime of the year it is when the gentle race of flowers dance over the teeming earth in the gayest guise and profusion.

McClain [1] quoted the 1841 writing of the Englishman, Mr. Oliver:

> A few miles further on we entered on a branch of Looking Glass Prairie, Bond County (Illinois), where long reaches of green undulating prairie stretched away until they became lost in the haze of distance; and, within a few hours of sunset, we emerged from a grove on the shore of one which lay stretched out before us like an ocean. In the direction which the track we were following took, we could just distinguish the forest like a low bank of cloud, whilst on our right the prairie stretched away, one vast plain, uninterrupted by tree or bush, as far as the eye could reach.

Dr. A.W. Herre [quoted in [3] p. 74] was particularly eloquent in his writings about the beauty of the prairie of yesteryears, and he sadly comments on its wholesale destruction by the 1890s:

> One of the most marvelous sights of my whole life, unsurpassed in my travels in nearly all parts of the world, was that of the prairie in spring. Unfading are memories of that waving

> rippling sea of lavender when the wild sweet William, a species of Phlox two or three feet in height was in full flower. It stretched away in the distance farther than the eye could reach. As the sea of phlox faded it was succeeded by another marvelous flower bed of nature's planting, and this in turn by others until mid-summer was reached. Then the great coarse perennials belonging to the Compositae dominated and instead of a single mass of color there was a vast garden of purple cone flowers, black-eyed susan's, rosinweeds, blazing stars, aster, goldenrods, and others...Every spring and fall the prairie was covered with water so that the whole country side was a great lake...All day long swarms of water birds filled the air, and far in the night their cries sounded overhead. At the first gleam of dawn vast flights of ducks dashed to and fro and great flocks of wild geese sped swiftly across the sky...The advent of tile drainage early in the (18)80's completed the transformation of the prairie into ordinary farm land and brought in many more people. Of course the ducks and geese stopped coming for there was neither water nor food to attract them. Migratory flocks of snipe and plover continued to come for a couple of decades, but their numbers had dwindled to a mere trickle when I left Illinois in 1900. The crayfish and bullfrogs disappeared in a hurry, and the prairie chickens were destroyed by the combined efforts of the plow and the shot gun. I returned to the region several summers during the (18)90's, but the prairie as such had disappeared, and of course, its characteristic life with it. What a pity that some of it could not have been preserved, so that those born later might enjoy its beauty also. Now it is merely flat unending corn fields, and moderns may look on this article as only the iridescent childish remance of an old man.

In recent years an appreciation for prairie has been shown by the extensive effort expended by individuals and government agencies to restore prairies. The Farm Bill 2002, Grassland Reserve Program, by USDA has resulted in major restoration by farmers to set aside marginal farmland. Unfortunately, some farmers have elected to plant alien grasses, which requires less expensive seed.

Tallgrass prairies need not be restricted to the Midwest. It is relatively easy to grow prairie in the East, where before habitation prairie ecosystems were somewhat rare because of largely unending forest. However, many "prairie" species can be found in mountain forest openings where human disturbance and introduction of Eurasian species has been minimal. Certain barrens in the East, such as one near State College, Pennsylvania, and the Serpentine Barrens on the eastern Pennsylvania–Maryland border, harbor many prairie species. The Serpentine Barrens have areas of shortgrass mixed with tallgrass, such as *Schizachyrium scoparium*, *Andropogon gerardii*, *Sorghastrum nutans*, and *Sporobolus heterolepis*, as well as several forbs. In the past there was a history of burns in the Serpentine Barrens. Except for a recent prescribed burn that became a wildfire, the remainder of the Serpentine Barrens is in a stage of succession into greenbrier, pitch pine, and red cedar. Volunteers of The Nature Conservancy are clearing areas of trees to promote grasslands. Otherwise, populated areas in the East are almost totally overrun with Eurasian species. Native wetland species often persist in the East, possibly because exotic wetland species have not been imported for agriculture and horticultural landscaping. However, the explosive growth of a wetland species, purple loosestrife (*Lythrum salicaria*), in the East demonstrates the hazards of unrestricted horticultural imports.

Most of the public considers "prairie" simply as a sea of grass. Such a view is far from accurate. There is a diversity of prairie ecotypes displaying a mixture of grasses and forbs (non-grass flowering plants). Depending on soil moisture, soil type, partial shade, local climate, and degree of disturbance, one prairie may have no resemblance to another in terms of plant species and species prevalence. The variance of prairie types adds an exciting dimension to their study and appreciation. An important key to prairie restoration is attention to the particular ecosystem outlined by the foregoing factors. Obviously, it is a total waste of time to seed a dry sandy area with wetland species or vice versa. A more detailed discussion of planting strategies is outlined in the Chap. 5 and Sect. 1.3 sections below.

As suggested by many experts, residential lawns should be supplanted with native species that normally grow wild in the region. Blue-grass lawns are a carryover from our northern European origins. Blue-grass lawns are an unnatural creation in the Americas. Enormous resources of water, fertilizer, herbicides, gasoline, lawn mowing equipment, and time are required to maintain our "English lawns." Maintenance of lawns in the likes of Tucson or Los Angeles is the ultimate in lavish environmental wastefulness, especially because of the scarcity of water in those arid climes. Prairie is a difficult sell in most subdivisions, but a small and growing minority is pleased with the results. Informing city fathers and neighbors before planting a prairie is recommended. And, one should organize attractive native plants in a manner that resembles a landscaping plan. Arranging areas of monoculture is one strategy that gives the impression of a suburban garden. A few years ago I planted beds of prairie plants for an acquaintance along Market Street in Camp Hill, Pennsylvania. Her lawn situated along a major thoroughfare attracts favorable notice.

1.2 Prairie Origins

A review by Axelrod [4] offers convincing evidence that an extensive prairie ecosystem is a recent phenomenon being maintained principally by fire during the time before settlement by European immigrants.

The prairies had their origin from the spread of grassy patches from open parkland or savannas in earlier times. In a warmer era from about 20 to 10 million years ago, conifers and hardwoods were found in river bottoms with grassy patches or open parkland in the intervals between rivers. Fossils of crocodiles and large tortoises indicated a warmer climate. As the climate became more arid 7–5 million years ago, the grasslands spread more into the savanna intervals. Nevertheless, at this time there was no extensive spread of prairie. A subsequent cooling trend led to more conifers, such as spruce and larch, with the parklands persisting until the end of the glacial period about 10,000 years ago. After this time, large expanses of prairie formed between river drainages, while river bottoms developed into deciduous forests.

The recent rise of the prairie agrees with the lack of speciation of prairie plants; that is, most prairie species are found in woodland borders. For example, many prairie plants can be found in the East in areas untouched by habitation, such as mountain openings. There are very few plant species endemic to prairies, indicating insufficient time to allow for the evolution of grass and forbs (flowering plants). Similarly, most bird and insect species are found elsewhere. Diverse relict trees show the former occurrence of extensive forest over the entire prairie region.

Axelrod [4] and many other authorities observed that these large expanses of prairies are unstable, particularly in the tallgrass region. After European immigrants populated the prairies, extensive fires were terminated, and existing prairies have been rapidly invaded by forest. Previously, for various reasons, Native Americans burned the prairies as often as every few years. Lightning may have played a minor role in setting prairie fires. This author personally witnessed a grass fire caused by lightning in eastern Wyoming; however, it was doubtful that it lasted long in the resulting downpour. It is this burn regimen that undoubtedly caused the major expansion of prairies into savannas after the last Ice Age. Other factors are involved, such as drying periods to permit the accumulation of flammable grasses, as well as flat plains over which the fire spread uninterrupted. Grazing animals may have also contributed to the demise of woody plants to a lesser extent.

1.3 Ecosystem Diversity

The focus of this book is the restoration of tallgrass prairies with attendant forbs commonly found in the north-central Midwest. As described below, it is entirely feasible to establish prairie ecosystems in the East in full sun or partial shade.

Prairie ecosystems are readily divided into the broad categories of prairies composed of tallgrass and those consisting of shortgrass. This division is mainly based on the annual rainfall, with regions of higher soil moisture being able to support tallgrass species. Within each of these categories there is considerable regional variation in species composition. However, the variability is gradual over the entire prairie ecosystem, so that one will find shortgrass species in tallgrass areas and vice versa. However, there are numerous species uniquely regional. Even in the tallgrass region alone you would find considerable similarity as well as variability between the species found in Texas from those in North Dakota. Even within a single species there are regional differences that undoubtedly have a genetic basis [5]. *Solidago altissima* was shown to change into regional ecotypes in Europe within only 300 years after its introduction (see discussion in Sect. 2.126 on *Solidago canadensis*), which may argue for an adaptation mechanism already present within the genome. For this reason, there is a variance of opinion regarding the selection of seed sources for restoration. Some experts recommend a 15-mile rule regarding seed collection for a specific site. The Illinois Nature Preserves Commission suggests a 100- to 200-mile radius within the same Illinois Natural Division, particularly on an east–west orientation.

The former range of tallgrass prairie was found from western Indiana west through Illinois and Iowa to central Nebraska. These prairies also stretched north into southern Wisconsin, southern and western Minnesota, and eastern South Dakota and North Dakota to southern Canada. Tallgrass reached south into northwestern

Missouri, eastern Kansas, and portions of east-central Oklahoma to parts of Texas roughly from Dallas south to San Antonio [1]. There were scattered areas of tallgrass prairie in southern Michigan, Ohio west of Columbus, central and western Kentucky, and east-central Arkansas. Even western New York and western Pennsylvania have small original prairie sites. Also, there were isolated pockets of tallgrass prairie surrounded by shortgrass prairie in eastern Colorado, western Oklahoma, and New Mexico.

Formerly, the East was forested, which did not permit the extensive spread of prairie. When pioneers cleared the land to plant crops, they introduced European plants familiar to them. Thus, full-sun/partial shade plants in the East are almost entirely of European and Asian origin. As mentioned previously, undisturbed openings in Eastern forests possess native plants that could be characterized as prairie. Fully 90% of the plants described here can be found in the East, and the remaining 10% easily can be grown in the East.

In prairie ecosystems there is considerable species variability as a consequence of soil composition and moisture. Often "dry" soils are composed of well-drained sand or gravel, but some species are simply attracted to the friability of sandy/gravelly soils. Surprisingly, sand and black prairie loam will support many of the same plants because both soils are loose and friable. On the other end of the spectrum there are truly wet prairies growing in standing water or water-saturated soils. Thus, the species composition of a wet prairie is virtually totally different from species found in a dry prairie. In comparison, it is hardly conceivable that both are categorized as "prairies."

In the past, considerable expanses of savanna existed in the Midwest where fire-resistant oak and hickory were separated far enough apart to permit prairie to grow in the interstices. Savannas were posited to be located in the areas between open prairie and forested gullies and stream drainages. It is likely that savannas were not subjected to frequent and hot windblown prairie fires resembling those that swept the flat plains. Thus, the savannas had a nearby source of tree seeds (acorns and hickory nuts) arising from the forested drainage areas, and savannas were exposed to somewhat less fire damage. Experts agree that the savanna ecosystem may be even more endangered than prairie. Occasionally, old savanna oaks with spreading branches can be seen decaying in the shade of a thicket of younger trees (Fig. 1.2).

The understory forbs and grasses that grew in the partial shade of these trees were both different from as well as the same as species found in full-sun prairie. For example, in savanna you are likely to find such species as *Dodecatheon meadia*, *Prenanthes* sp., *Elymus villosus*, *Geranium maculatum*, *Aureolaria grandiflora*, *Lespedeza violacea*, *Thaspium trifoliatum*, *Anemone virginiana*, *Agastache scrophulariifolia*, *Asclepias purpurascens*, *Arnoglossum atriplicifolium*, *Carex davisii*, other *Carex* spp., *Gentiana alba*, *Heuchera richardsonii*, *Lactuca* sp., *Heracleum maximum* (moist), *Polemonium reptans*, *Polygonatum biflorum*, *Rudbeckia triloba*, *Sanguinaria canadensis*, *Echinacea purpurea*, *Taenidia integerrima*, *Cypripedium parviflorum*, *Trillium recurvatum*, *Campanulastrum americanum*, *Chelone glabra*, *Hypericum ascyron*, *Glyceria striata*, *Bromus* sp., *Panicum/Dichantelium* sp., *Lobelia cardinalis*, *Phlox divaricata*, *Lilium superbum*, *Eryngium yuccifolium*, and others.

Fig. 1.2 Savanna oak invaded by young trees. In a decade the tree died from loss of limbs

Soil disturbance also plays a surprisingly significant role. For the state of Illinois, Taft et al. [6] have classified species into 11 levels of tolerance to soil disturbance, which they called the Coefficient of Conservatism (CC); see Appendix 1 and Part II (Chaps. 2–4), List of Species. The values range from a rating of 0, indicating weedy species growing in disturbed soil, to 10, indicating undisturbed situation or a high degree of "fidelity to habitat integrity." Thus, *Amorpha canescens* rated at a CC of 8, *Dalea candida* at CC 9, *Sporobolus heterolepis* at CC 9, and *Gentiana puberulenta* at CC 9 are indicators of undisturbed high-quality prairie. Such high-quality prairie plants are particularly difficult to establish. A rating of CC 10 is usually reserved for rare, threatened, or endangered species. Often special methods are required to propagate conservative plants. Initially, restorations will comprise species mainly with CC values of 0 to 7. Later on, more conservative plants may appear from earlier seed distribution, or preferably you may choose to scatter seed of conservative species after the prairie is established. CC values have been developed for a few states other than Illinois.

The environmental/developmental factors discussed above lead to a diverse variance of species between different prairies. Only when all these factors are close to identical can one expect a reasonable likelihood of finding two similar prairies.

As examples, Figs. 1.3–1.6 show the aspects of some different prairies growing in ecosystems of varying available moisture and soil type. Regarding tolerance to soil disturbance, there are even differences in early succession and late succession prairies, as already discussed. One can hardly conceive that these various ecotypes are all classified as tallgrass prairies. All the prairies shown here are typical of ecosystems found within a 50-mile radius of Peoria, Illinois. Because of Peoria's proximity to the Illinois River, the ecology ranges from sand dunes to wetlands, to gravel-laden river bluffs and graded forests, and to black prairie loam. Interestingly, the Illinois River was the pre-Ice Age bed of the Mississippi River south of the "Great Bend" located near the Hennepin Canal. Consequently, it left a deep river valley and quantities of sand and gravel. Over the millennia the sand dunes migrated with the prevailing wind a few miles east of the river.

Figure 1.3 represents a black prairie loam found east and west of the Illinois River, where in former times there were areas of seasonal standing water. Figure 1.4 shows an area with good drainage of cleared "forest soil." Shown in Fig. 1.5 is an Illinois River bluff composed mainly of gravel admixed with sand. Figure 1.6 shows a "created dry" prairie of imported sand, which represents a typical ecosystem of sand dune prairies found east of the Illinois River.

Fig.1.3 Wet-mesic prairie on black prairie soil (silt-loam) at the Rock Island Trail Nature Preserve, Illinois

Fig. 1.4 Mesic prairie on author's land along Switzer Rd., Brimfield, Illinois

Fig. 1.5 Dry-mesic hillside prairie found on gravelly-sandy soil. Location: Manito Prairie Nature Preserve, Illinois (Photo: Debra Grim)

Fig. 1.6 Sand prairie, usually rated as "dry"

1.4 Prairies Attract Wildlife

Prairies attract a wide variety of birds, mammals, insects, and even amphibians. In former times the springtime chorus of frogs and waterfowl in wet prairies was legendary. It was said that "Froggy Prairie" in Illinois earned its name from the fact that the teacher in the prairie log schoolhouse had to speak "at the top of his voice" to be heard over the noise of the frogs [1]. Because most wetlands in the Midwest have been drained for agriculture, Froggy Prairie and others like it have been extirpated from memory.

Except for wildlife preserves, the large grazing animals, bison and elk, have been eliminated from the few remaining tallgrass prairies. Although one may think of bison as formerly inhabiting prairies from the Midwest to the Rockies, a larger longer-legged eastern bison ranged from New York to Georgia [7]. The eastern bison migrated from summer pastures along Lake Erie to winter in mountainous areas and valleys as far eastward as the Susquehanna River in Pennsylvania. The eastern bison is now extinct. In Pennsylvania, the last individual was killed in 1801. It is said that the eastern bison preferred haven in dense forests but could be found sunning themselves in open fields in the winter. Thus, there is a certain implication of the former presence of open grazing meadows or fields in the East. It would have been of considerable interest to examine the native plant species content of these bison grazing areas. Similar to the bison, the eastern elk was extirpated by the late 1870s [7] but has been reintroduced from western stock.

Smaller mammals find refuge even in small plots of tallgrass prairie located in the midst of large expanses of corn and soybean fields, such as coyote, deer, badger, and a variety of rodents. In the East, woodchuck and deer find a home there. The presence of deer can be a mixed benefit. In small numbers deer are not especially detrimental to plant life, but in large numbers they selectively browse certain species, which eventually decreases diversity. Before settlement by the white man, the deer population reportedly was much reduced because of the presence of predators and the type of browse. Ecosystem fragmentation has been cited as a reason for deer overpopulation. This author encourages deer hunters on his property, which is an attitude shared by many native plant enthusiasts. Large numbers of deer are akin to a herd of goats.

Open prairies specifically attract certain birds, such as the increasingly rare dickcissel, meadowlark, prairie chicken (very rare), Henslow's sparrow (very rare), shrike, kingbird, and bobolink. In wet prairies in more northwestern prairie states the yellow-headed blackbird is found, as well as many waterbirds. More common birds, such as red-winged blackbird, bobwhite quail, killdeer plover, goldfinch, bluebird, and a number of sparrow species (vesper, lark, field, grasshopper, savannah, swamp, and song) regularly visit or nest in the prairie. Ruby-throated hummingbirds are attracted to certain species of flowers, and their presence is an occasional welcome sight. In a search for prey, hawks such as red-tailed hawk, sparrow hawk/kestrel, and marsh hawk/harrier are occasionally seen in open grasslands.

Snakes are often seen in the tallgrass, especially bull snakes, lined snakes, and the docile red-bellied snake. Prairie fires can decimate snakes when they emerge from hibernation. Dead snakes after late burns are a reminder that snakes are perhaps among the more susceptible to fire. Eastern box turtles would be susceptible to fires in their eastern range.

Perhaps insects and arachnids are the least obvious inhabitants of the prairie. A survey of remnant prairies and savannas in the Chicago region by Panzer et al. [8] documented 2,424 insect species, of which 412 conservative species were largely associated with remnants rather than degraded fields. The prairie/savanna-inhabiting insects were mainly grasshoppers, walking sticks, katydids, leaf hoppers, frog hoppers, plant hoppers, butterflies, moths, and beetles. Panzer [8] postulated that there could be as many as 2,000 conservative insects narrowly associated with remnant prairies and savannas. Although there are diverse species of arachnids present in prairie, the most obvious are large garden spiders. Even after years of barging through their webs, a garden spider has not bitten me. Management of prairies by burning undoubtedly is detrimental to certain insect and arachnid species. Thus, the Illinois Nature Preserves Commission makes certain that only portions of prairie tracts are managed by fire. The burning of portions is rotated on an annual or semiannual basis.

Tallamy [9] writes that insects and arachnids are the keystone to the maintenance of many vertebrates. Birds, amphibians, reptiles, and fish depend on insects and arachnids for their livelihood. In particular, bird nestlings need high concentrations of protein and fat provided by insects and arachnids. About 96% of bird species rely on such food sources to rear their young. Tallamy's studies [9] showed that our native insects are evolved to feed on native plants, including native trees. In other words, they have adapted over the millennia to digest the secondary metabolites produced by native plants. Certain insects feed on one specific native species. There are "generalist" insects that feed on a number of plants, but even these insects prefer natives. To add to the insect's problems, importers of exotic alien plants often select for insect resistance. According to Tallamy [9], native plants generate 35 times more insect biomass used by birds than exotic alien plants. Development and agriculture continue to decimate much of our native plants. More than 62,500 square miles in the United States have been converted to sterile suburban lawns, which are planted with alien exotics, fertilized, and treated with insecticides and herbicides. Your "Kentucky" blue-grass lawn originates from Eurasia, and your lawn supports the alien Japanese beetle grub, which feeds on Eurasian grass roots. As a result, there has been a 50% decline in the population of neo-tropical migrant birds since 1966 [9]. Admittedly, it is frustrating to witness insect predation of some of your hard-to-grow native plants, but it is a sign of a healthy ecosystem capable of supporting more life. Unfortunately, there is an overall dearth of native plants as well as serious ecological fragmentation that will not ever support historical levels of wildlife in the United States.

The wildlife species discussed here are incomplete, comprising those most commonly found on average prairies. For a more complete list, consult Packard and Mutel [10].

Part II
Native Prairie Species

Although there are a few annuals and biennials found in prairies, the majority of prairie species are perennial; some live for many years. This persistence answers the oft-asked question about prairie burns: "Don't you have to replant the burned area?" There is an incredible biomass invested in the root system of perennials (two thirds of the entire biomass), explaining why it takes so long to see obvious evidence of prairie species after planting (at least 3 years for tallgrass and most forbs). Some of the drought-resistant *Silphium* species have taproots 8 ft long. Most of the tallgrasses have roots 4–6 ft deep. And, *Baptisia alba* v. *microphylla* has been noted to have a root system about 10 ft deep. Mycorrhizae, the soil fungi, associate symbiotically with plant roots, providing them with essential nutrients such as phosphorus. Thus, your restored prairie has interesting features that will be revealed to you. Because of the extensive root depth of certain species, it not recommended to transplant many without further information. The use of seeds is the preferred method of restoration. Interestingly, I have grown prairie species in soil only 2 ft deep underlain with seemingly impenetrable shale. One wonders if the roots find their way through cracks in the shale layers or grow laterally.

The species outlined below give specific information about each plant, including instructions for restoration. The selected plants are those most commonly found in mesic prairies centered in Illinois. Wetland and sand-loving species are somewhat underrepresented. Also included are some less common and savanna-preferring plants that add interest to prairie ecosystems. All plants listed here grow in the tallgrass region of the Midwest, and additionally the author has found that they will thrive in the Mid-Atlantic region of the East. In the East many species range further north and south than the Mid-Atlantic region. Those species denoted by an asterisk * have been found to occur in both the East and the Midwest, including adventive and escaped species. Fully 90% of so-called prairie plants occur in the East. In general, most plants described here occur from the Northeast to the plains east of the Rocky Mountains, often including the lower Canadian provinces, and range to the southern border. Very often the ranges of plants exclude their occurrence in the Far West, South Carolina, Florida, and Louisiana.

Despite the claim that the scientific name is "the one and only" nomenclature for a plant, authorities usually have lavished many scientific names for a single species. In instances where there has been a recent name change, a formerly used scientific nomenclature is listed here. Consult http://plants.usda.gov/java/nameSearch for a complete listing of current and former scientific nomenclatures.

Treatment of seeds for germination of individual species is often based on the recommendations of Prairie Moon Nursery (11). As recommended by Prairie Moon Nursery (11), a cool-moist stratification (40°F) for 2 months is the most common method to trigger germination (usually seeds mixed with moist sand, not overly wet). Professor Deno has experimentally determined precise germination methods for selected species, and these data are given if available (12). Deno points out that there often are germination inhibitors in seeds that must be destroyed, usually by a 3-month cool-moist treatment (40°F) used by him or simply dry storage. Another common factor is a requirement for light. Deno has studied the use of the fungal-produced germination hormone, gibberellic acid-3, but these data are given only when it is the best alternative for triggering germination. Optimal germination directions are given for each species outlined below. Also, see Sect. 5.6 for a summary of the various methods used to cause germination of specific species. Appendix 1 gives germination codes for additional species not detailed in this book.

The Coefficient of Conservatism of different species, based on the work of Taft et al. (6), indicates tolerance to disturbance or fidelity to habitat integrity. The highest value of 10 indicates the highest order of plant succession or intolerance to disturbance; a value of 0 indicates a very weedy species that will thrive on disturbed soil.

Detailed images of the individual species, as well as the seed-bearing plants with their seeds, are available as herbarium scans on the Springer website. The herbarium images are available in color as scans of fresh samples (200 dpi), as well as gray-scale scans of either dry-pressed samples or fresh samples (300 dpi). Further information about individual species can be found at http://plants.usda.gov/java/nameSearch on the web, including species and varietal distribution.

This chapter is divided into three categories of species preference for soil moisture: Mesic soil and adaptive species (Chap. 2); Dry-mesic to dry soil-preferring species (Chap. 3); Wet-mesic to wet soil-preferring species (Chap. 4). The species are arranged in alphabetical order in each section rather than being ordered by their botanical classification.

Chapter 2
Mesic Soil and Adaptive Species

Categorizing species by their preference for soil moisture is the recommended method to ensure that the restoration proceeds successfully. The category of species preferring mesic soil (medium soil moisture) is the most often encountered. Adaptive species are included as they are often found in soils ranging from very moist to dry soils.

In addition to soil moisture types, one must be aware of sunlight requirements, such as full sun or partial shade. These conditions are described for each individual species.

2.1 **Allium canadense* v. *canadense* (Liliaceae Family), Wild or Meadow Garlic

Wild garlic, *Allium canadense* v. *candense* (20–50 cm in height), is found in prairies, savannas, and woodlands in soils ranging from wet mesic to dry mesic (illustration on page 102). When wild garlic is found in dry-mesic soils, it is usually growing in shade. This plant is found from Maine to North Dakota and south to the U.S. border, as well as in Quebec and Ontario. There are five other varieties with more restrictive distribution. Flowers and bulblets develop simultaneously on the flower head. Restoration can be accomplished by collecting bulblets from the flowering heads in mid- to late June, followed by immediate scattering of bulbs or shallow planting on tilled soil. Storage for a period of a month or more in plastic bags in refrigeration, not freezing, is acceptable; however, in nature the bulblets fall to the ground when mature, arguing for immediate planting. This author has not observed sprouting of the bulblets in midsummer; thus, they usually lie dormant until spring. In open prairies wild garlic thrives on black prairie loam, but survives satisfactorily in other soils if not too dry. Wild garlic is also found in the partial shade of woodlands. According to Taft et al. [6], the Coefficient of Conservatism is only 2 (also see Appendix 1); thus, this plant is likely to be found in more disturbed areas decreasing in abundance as the prairie matures. *A. canadense* starts to senesce after blooming, so it will disappear during the heat of midsummer only to appear again in the spring. This species

H.W. Gardner, *Tallgrass Prairie Restoration in the Midwestern and Eastern United States: A Hands-On Guide*, DOI 10.1007/978-1-4419-7427-3_2,

should not be confused with *Allium vineale*, a similar Eurasian import. According to Britton and Brown [13], its pink or white flowers distinguish *A. canadense* as well as its bulbous root covered with a fibrous-reticulated outer coat, whereas *A. vineale* has green or purple flowers with a bulbous root covered with a membranous coat. In the East this Eurasian species is very aggressive in disturbed soil and lawns but substantially decreases as the prairie plants become established.

2.2 **Allium cernuum* v. *cernuum* (Liliaceae Family), Nodding Wild Onion

It is said that the Indians had two meanings for the word "chicago"; one was "smelly place" and the other was *Allium cernuum* v. *cernuum* (illustration on page 102). By little coincidence the Chicago area once had a considerable population of nodding wild onion. It is a species with a fairly high Coefficient of Conservatism of 7; thus, it is found largely in undisturbed prairies ranging from New York to Washington and south, except the Southwest and the extreme Southeast. The variety *cernuum* is also found in most areas of southern Canada from Ontario to British Columbia. There are two other varieties, one in the Southwest and another ranging from the Northwest to the Southwest. Although it prefers wet-mesic to dry-mesic soils, it is fairly tolerant of midsummer drought. *Allium cernuum* (height, 30–50 cm) can be distinguished by the pronounced nodding of the flower as compared with the very similar native onion, *Allium stellatum*, which has an upright flower. *Allium cernuum* v. *cernuum* generally flowers in mid- to late July, 1–2 weeks earlier than *A. stellatum*. Both *Allium* species are attractive additions to a prairie restoration. The entire seed-head can be collected from early to late September after the majority of seed pods split open to reveal the black seeds (about 2 mm in diameter). After drying the seed-head, the seeds drop free. Full recovery plus chaff can be obtained by rubbing over a sieve made of quarter-inch hardware cloth. The seeds respond to fall planting, and the plants thrive under cultivation. According to experimental data [12], seeds germinate at 70°F (50–65% germination with an induction time of 3 days at a rate of 0.7% per day) using both fresh seed and seed dry-stored for 6 months at 70 or 40°F. Moist temperature cycles gave a higher total percentage germination; that is, fresh seeds germinated after 3 months each of moist cycles of 40°F (0–10%), 70°F (50–55%), 40°F (20–25%), and 70°F (20–25%). Seeds dry stored for 6 months gave a total germination of 87% through one cycle each of 40 and 70°F each of 3 months duration.

2.3 *Allium stellatum* (Liliaceae Family), Prairie Onion

Except for the straight neck of the flower stem and the leaves, *Allium stellatum* generally resembles *A. cernuum*, both in height (20–45 cm) and appearance of the flower and seed (illustration on page 102). Prairie onion is found in most of the Midwest from Michigan to North Dakota and south to other states, as well as Ontario to Saskatchewan in

Canada. *A. stellatum* does tend to have a fuller flower head, and it prefers somewhat drier conditions of mesic to dry soil. The leaves of *A. stellatum* are thick, hard, and rounded on the back, as well as channeled, whereas *A. cernuum* leaves are thin, soft, and flat. This onion gives attractive blooms in early August, and seed-heads are best collected from mid-September to early October. Seed-heads are clipped when the capsules open to reveal the black seeds. See Sect. 2.2 for seed recovery. Seeds require a 2-month cold-moist stratification for germination [11]. In agreement with the assignment of the highest possible Coefficient of Conservatism of 10, this species is very difficult to establish in prairies. By contrast, it can be easily cultivated by row cropping or in weeded beds. Evidently, prairie onion prefers mature ecosystems and/or does not tolerate much competition from other plants.

2.4 *Amorpha canescens* (Fabaceae Family), Lead Plant

Amorpha canescens is named lead plant because of the somewhat leaden hue of the leaflets (illustration on page 102). Lead plant ranges through Michigan, Indiana, Illinois, Wisconsin, and west of the Mississippi River to Montana and south to the border, as well as in Ontario and Manitoba in Canada. Purplish flowers with yellow stamens appear in early July. Lead plant is considered a conservative (Coefficient of Conservatism, 8), that is, an indicator of pristine prairies. *A. canescens* is one of the few woody prairie plants that leaf out in the spring from the original stem, reaching a height up to 1 m; however, after a prairie burn it resprouts readily from the root. In my opinion repeated annual fires are detrimental to lead plant, as it loses height in competitive prairie ecosystems. Although lead plant prefers a range of mesic to dry soils, it is difficult to restore in rich mesic soils as these ecosystems are usually original prairies requiring many years to develop. However, it is relatively easy to propagate by simply sowing on poor, dry soil in either full sun or partial shade. Because poor soil affords little competition from other plants, it is the probable reason for its ease of restoration in marginal soil. The seed is a tiny "bean" contained within a 4-mm-long tomentous gray pod. In mid-September to mid-October the gray pods can be stripped by hand from the seed stalk. The seed contained within the pod can be fall-planted without dehulling. Under these circumstances the seed may lie dormant during the next growing season only to germinate the following year(s). Because of the desirability of growing this rare plant, it is advisable to use the recommended seed treatment of pod/hull removal and 10 days of moist-cold stratification [11]. Vigorous scraping of the hulled seed over a window screen placed an inch over a hard surface will remove portions or the whole of the tomentous hull. Extraneous hull material, including the dehulled seed, will fall through the screen. The hull of lead plant seed is much more resistant to scraping compared to *Dalea* species; thus, you may opt to recover the remaining hulled seeds from the top of the screen. In my hands, satisfactory germination was obtained with hulled seed that was previously subjected to scraping. To restore lead plant in rich soil, it is recommended that plants are started in a nursery, growing in 10- to 20-cm-tall pots, as early as February. April plantings in pots do well if grown in a sheltered sunny place free from frost. Spread dehulled seeds on top of soil and barely cover with a thin layer of sand. If peat pots are used, there is the danger of mold development, which causes damping-off of seedlings. The peat pots can be premoistened for a period of time to allow molding to run its course or treated with fungicide. Plastic pots avoid the mold problem. After all danger of frost the plantlets can be placed outside; initial weeding around the seedlings is recommended to reduce competition. Peat pots often attract digging by wildlife (perhaps squirrel or raccoon), so it is advisable to remove the peat pots before planting or cover the plants with a wire cage.

2.5 ***Amsonia tabernaemontana v. salicifolia (Apocynaceae Family), Amsonia or Blue Dogbane**

Amsonia tabernaemontana v. *salicifolia*, amsonia or blue dogbane (height, 0.4–1 m), is generally found in rich woodlands but it will grow in open, wet-mesic to dry-mesic prairies. Variety *salicifolia* is found from Virginia west to Illinois, Missouri, and Kansas and south, except Florida. Comparatively, variety *tabernaemontana* additionally extends its range from Virginia north to Massachusetts and New York. Blue dogbane often selectively grows well on wet-mesic bottoms; however, it is surprisingly tolerant of drought conditions, as well as dry-mesic soil. In mid-May, the attractive blue flowers make a good addition to a prairie landscape. The plants are fairly long-lived, being only partly damaged by intense prairie fires. The plant will recover, as shown by the photograph taken following a March prairie fire (see page 102, including inset showing close-up of flower). Its Coefficient of Conservatism is 6. Amsonia seems to have variance in forms. Among individual plants leaves vary in size and width; however, some of this variance may be the result of the amount of sunlight available. One distinct variety (v. *salicifolia*), shown here, differs by having lanceolate leaves 1–3 cm broad with glaucous (whitish bloom) undersides versus the typical variety, v. *tabernaemontana*, with elliptic-oblong to ovate leaves 3–6 cm broad, and the plant of v. *salicifolia* usually is half the height. Another variety (v. *gattingeri*), restricted to Kentucky south to Alabama, is also known. Clusters of cylindrical pods (9 cm × 3 mm) can be collected when the pods begin to split open, from early September to early October, revealing the brown cylindrical seeds (about 6 by 1–2 mm). Fall planting is a generic method that is successful with the author. According to Deno [12], germination is 29% in 3–7 days after a cycle of 3 months of moist 40°F followed by 3 months at 70°F. When sown at 70°F, there was no germination over three cycles of 3 months.

2.6 ***Andropogon gerardii (Poaceae Family), Big Bluestem Grass**

Andropogon gerardii, big bluestem grass (height with seed-head, 1–2.3 m), is a major species of the tallgrass prairie (illustration on page 102). Big bluestem is found in southern Canada and all the U.S. except the far West. In a dense stand of this magnificent grass, it is easy to understand how a pioneer could get lost in a flat prairie. It thrives in full sun or savannas in wet-mesic to dry-mesic soils. Although it has a medium Coefficient of Conservatism of 5, it is easy to establish. This author successfully propagated it by sowing it either spring or fall. It is recommended to sow seed after killing perennial weeds with glyphosate (Round-up herbicide). Coincidentally, the patent for Round-up has run its course, thus less expensive brands of glyphosate (isopropylamine salt) are available. Sowing on tilled soil seems slightly more effective than untilled soil, but tilling gives increased danger of bringing weed seeds to the surface. Spring planting with a range drill after glyphosate treatment was highly successful. Spring planting with a grain crop is another option. Expect to wait 2–5 years before this grass becomes notable. It spends years growing a root system. The seeds do not require any treatment [11]. Germination experiments [12] showed that seeds germinated 76% in 3–6 days at 70°F. A temperature cycle of 3 months of moist 40°F followed by 3 months at 70°F gave 35% germination in 2–11 days. Seeds can be collected during October by stripping the seed-head. The seed-head segments into pieces of about 7–14 mm, which are largely composed of extraneous material such as the awn. The seeds should be spread out on a dry surface or screens to dry. Many nurseries, such as the Illinois Department of Natural Resources, process the impure seeds to obtain dark seeds about 6 mm by less than 1 mm. Seed purification is totally unnecessary for the amateur not interested in commercializing the seed. A dense planting of big bluestem grass can inhibit the restoration of other species, and 15 years or more is required for ecological equilibrium. In my experience, the species that initially coexist most readily with big bluestem are *Lespedeza capitata*, *Silphium laciniatum*, and *Baptisia alba* v. *macrophylla*. For a more rapid achievement of ecological equilibrium with forbs, a light planting of *A. gerardii* is recommended, with more emphasis on shortgrasses, such as *Schizachyrium scoparium*.

2.7 *Andropogon scoparius* see *Schizachyrium scoparium*

2.8 **Anemone canadensis* (Ranunculaceae Family), Canada Anemone

Anemone canadensis (height, 30–45 cm), a plant preferring wet-mesic to mesic soils, is found in savannas (partial shade) or open prairies (illustration on page 103). It is native from Maine west to Montana and in the south from Virginia west to Kansas and New Mexico. It is found in all the southern provinces of Canada, including the Northwest Territory. Canada anemone, with its umbrella-like leaf and white flowers, provides an attractive sight in mid-May. It is considered a mid-conservative in prairie ecosystems. After being collected in mid-July to late August, the seeds should be given alternating 2-month periods of cool-moist, warm-moist, and cool-moist for germination [11]. Fall or spring planting may require a second season or longer to germinate. In my hands, seeds planted in spring were completely dormant the first and second season. In late summer, the plot was sprayed with glyphosate to control weeds. Young plants were first noted in July of the third year, which fulfilled the requirement for a sequence of cool-moist, warm-moist, and cool-moist periods. Although it tries your patience to restore this plant by seed, it is far simpler to transplant whole plants. Transplanted individuals readily spread by root rhizomes, usually in dense patches. With little competition, Canada anemone can be rather aggressive; however, it does not seem to spread aggressively in established prairies. This species can be an excellent ground cover in partially shaded suburban gardens or lawns. Most home gardeners find that they need to control its spread after several years of growth.

2.9 ******Anemone virginiana* v. *virginiana* (Ranunculaceae Family), Tall Anemone or Thimbleweed

Anemone virginiana v. *virginiana*, tall anemone (height, 0.4–1 m), is usually found in mesic to dry-mesic soils where it is found from Maine west to North Dakota and Wyoming, and in the south from Georgia west to Colorado; it is also found in Quebec and Ontario (see illustration on page 103). Two other varieties are found in more northerly areas. Although tall anemone prefers savannas, it will grow on open prairies. A fairly inconspicuous flower appears in late June. In springtime the purplish-green leaves lend an attractive feature, especially if exposed to full sun. The seed-head, which resembles a thimble, can be collected from early September to mid-October when the seed-head begins to show signs of breaking or browning. This "thimble" breaks apart into fuzzy balls about 2–3 mm in diameter that tend to cling together. Diligently mixing the clinging mass with sand or other seeds permits their separation into individual seeds. According to Prairie Moon Nursery [11], seeds need no treatment for germination if planted early. Thus, a late fall planting or a spring planting of dry stored seeds will give satisfactory results. Tall anemone has been assigned a Coefficient of Conservatism of 4. This author has had success with fall planting seeds in beds for the purposes of seed production. Herbivores, probably deer, selectively browse the plant. Dog hair clippings spread around the plants discourage grazing.

A similar plant, **Anemone cylindrica*, is found in open prairies, and it can be distinguished by its cylindrical seed-head (1.7–3.4 cm long by 0.7–1.1 cm thick) compared to the elliptical or oblong seed-head of *A. virginiana* (2.3–3 cm long by 1.2–1.5 cm thick). Also, *A. cylindrica* leaf divisions are more wedge shaped and narrow, whereas *A. virginiana* has more ovate and broad divisions of the leaves, mostly convex at the base. Otherwise, these two species can be fairly similar in appearance. Similar to *A. virginiana*, *A. cylindrica* requires no treatment for seed germination [11]. Probably because of weed cover, seedlings were not noted the first season after planting, but appeared in late March of the following year.

2.10 **Apocynum cannabinum, Apocynum sibiricum* (Apocynaceae Family), Indian Hemp or Dogbane

Apocynum cannabinum, Indian hemp (height, 0.3–0.7 m), is often found in early succession prairies with wet-mesic to dry-mesic soils (illustration on page 103). This plant is found throughout the U.S. and most of Canada. Indian hemp blooms in late June to early July. It spreads readily by root rhizomes. Seeds are borne in typical cylindrical dogbane follicles (pods) about 4–20 cm by 2 mm wide. Seeds are retained in the pods into the late fall; thus, the seed heads can be harvested at convenient times. Cylindrical brown seeds are about 5 mm long, attached to fluffy hairs 0.8–3 cm long. Seeds need no treatment for germination. A Coefficient of Conservatism of only 2 is indicative of its weedy character. Indian hemp tends to decrease as a prairie matures in ecological succession. According to some authorities [13, 14], **A. sibiricum* is a separate species from *A. cannabinum*, and *A. sibiricum* can be distinguished by having a very short or no petiole, compared to the short petiole of *A. cannabinum* (0.2–1.3 cm). Also, *A. sibiricum* has parallel primary leaf veins forming broad angles with the midvein. However, Barkley et al. [15] point out that *A. cannabinum* is a highly variable species, which has overlapping characteristics with *A. sibiricum*; thus, they consider *A. sibiricum* unworthy of species designation. In fact, plants with descriptive characters of both species often can be found growing together.

2.11 **Arnoglossum atriplicifolium, Cacalia atriplicifolia* (Asteraceae Family), Pale Indian Plantain

Arnoglossum atriplicifolium, pale Indian plantain (height, 1–2 m or more), is primarily a savanna plant, but it is sometimes found in full sunlight (see illustration on page 103). It prefers mesic to dry-mesic soils in partial shade or mesic soils in full sun. Pale Indian plantain is distributed from Massachusetts west to Minnesota and Nebraska and south and east to all states except Texas. Small, white, rather inconspicuous flowers appear in early July. However, the smooth (upper side glabrous, lower side glaucous) lobed leaves and tall aspect of the plant are the main objects of interest. This plant is rated in the middle of the scale of the Coefficient of Conservatism. Seed harvest can occur when the seed-head becomes fluffy with white pappus in mid-September to early October. The seed-heads bearing multiple seeds should be clipped carefully to prevent seed loss and placed in a paper bag for subsequent processing. Seeds are also readily lost during windy periods. After drying the seed-heads a short time, the seeds plus their pappus can be easily shaken loose for collection. Because the fine pappus hairs readily disperse into the air and irritate the respiratory tract, wearing a facemask and preferably working outdoors is strongly recommended. The seed plus pappus is about 1 cm long and the black seed is about 4 mm long. The recommended seed treatment for germination is 2 months in cool-moist stratification [11].

2.12 **Arnoglossum plantagineum, Cacalia plantaginea/tuberosa* (Asteraceae Family), Prairie Indian Plantain

Except for the ovate leaves, *Arnoglossum plantagineum*, prairie Indian plantain (height, 0.6–2 m), appears similar to the preceding species, *A. atriplicifolia*. The photograph on page 103 shows the very early stage of blooming (see Springer internet herbarium illustration to view a mature flower). This species is found in wet-mesic to mesic prairies, especially black prairie loam. With few exceptions this plant is absent from the eastern coastal states. Its main habitat is from Ohio and Ontario west to Minnesota and South Dakota. From this northern limit, it ranges south to the border. Flowering occurs from mid-June to mid-July. The highest Coefficient of Conservatism of 10 indicates its scarcity and difficulty of restoration. Seed collection and treatment for germination is the same as recommended for the preceding *A. atriplicifolia*. And, as with *A. atriplicifolia*, avoid breathing the fine pappus hairs. This author succeeded in obtaining germination and growth with a fall planting in dry-mesic shale soil with persistent weeding and occasional irrigation during the growing season. My in-field restoration efforts appear to be unsuccessful thus far.

2.13 **Asclepias purpurascens* (Asclepiadaceae Family), Purple Milkweed

Asclepias purpurascens, purple milkweed (height, 0.6–1 m), prefers partial shade, such as found in savannas, but it does grow in full sun (illustration on page 103). Purple milkweed is found from Maine west to Ontario, Minnesota, and South Dakota and all states south of this limit, except South Carolina, Alabama, and Florida. Its striking purple color in mid-June justifies including it in restoration projects, particularly if some partial shade can be provided. With a Coefficient of Conservatism of 7 it is moderately difficult to establish, and it generally is found as occasional individual plants. However, rhizomes can spread this milkweed in a less aggressive manner than many other milkweed species. Mesic to dry-mesic soils are required, and as with other milkweeds, friable, well-drained soils are better. The seedpod can be collected in mid-September when the seeds are a mature brown in color. A good test for seed ripeness involves pinching the pod to determine if it easily splits open. The seedpods are relatively smooth as contrasted with the seedpod of the weedy common milkweed (**Asclepias syriaca*), the latter of which is tomentous and covered with short protrusions. One can strip away the seeds while tightly grasping the top of the cluster of silk (coma) after carefully removing it from the pod. In the event of premature opening of the pods or the processing of large quantities, the seed can be easily separated from the silk by rubbing across quarter-inch hardware cloth. However, to obtain good yields by this method, the pods should be removed first as the pods tend to entrap both seeds and silk. The seeds are flat brown ovals about 6 mm in length, and they require 2 months of cool-moist stratification or no treatment for germination [11]. Note: According to Prairie Moon Nursery [11], an alternative method for all the *Asclepias* species is no treatment for germination, which is not entirely correct for all *Asclepias* species.

2.14 *Asclepias sullivantii* (Asclepiadaceae Family), Sullivant's or Prairie Milkweed

Asclepias sullivantii, Sullivant's milkweed (height, 60 cm–1.2 m), is typically a Midwestern milkweed distributed in an area from Ohio and Ontario in the Northeast to North Dakota in the Northwest and south to Oklahoma and Arkansas (see illustration on page 103). Sullivant's milkweed is often mistaken for common milkweed, *A. syriaca* (not shown), but *A. sullivantii* has somewhat more showy flowers blooming in early July. Other comparisons of the two milkweeds clarify the differences. Sullivant's milkweed tends to have narrower leaves clasping the stem, and the leaves are glabrous on both sides, whereas common milkweed has a densely pubescent leaf underside. In addition, the seedpod of common milkweed is rough with protuberances. As with *A. purpurascens*, the seedpod of Sullivant's milkweed is relatively smooth. Sullivant's milkweed can be distinguished from purple milkweed by its habitat and by its narrower, clasping leaves. The seed-pods can be collected from late August to mid-October, and seeds can be obtained from the pods in the same manner as already described for *A. purpurascens*. The seeds are flat brown ovals about 8 mm in length. To germinate, the seeds require 2 months of cool-moist stratification or no treatment [11]. This author had good success with fall planting in glyphosate-treated beds. Possibly reflecting its fairly high Coefficient of Conservatism of 7, Sullivant's milkweed requires somewhat restricted conditions to thrive, that is, mesic to wet-mesic open prairie with friable soil, such as prairie loam. Although it will grow in mesic soil, more moisture is desirable. Because it readily spreads by root rhizomes, row cropping is nearly impossible for reasons of its tendency to migrate from the row.

The aforementioned **Asclepias syriaca* (common milkweed) is also a native prairie species, but restoration of this species would not be desirable. With a Coefficient of Conservatism of 0, common milkweed is considered a weed that usually invades without help, especially in recently restored prairies.

2.15 **Asclepias verticillata* (Asclepiadaceae Family), Horsetail or Whorled Milkweed

Asclepias verticillata, horsetail milkweed (height, 0.3–0.8 m), is a rather weedy plant (Coefficient of Conservatism, 1) that is found in more disturbed ecosystems, mainly on mesic to dry hillsides exposed to either full sun or partial shade (illustration on page 103). Horsetail milkweed is widely distributed in the U.S., except for the upper Northeast and the West. In Canada it is found in the lower mid-provinces. In areas of low competition horsetail milkweed spreads rapidly by rhizomes; however, it can be surprisingly sensitive to drought and competition from other plants. As a prairie matures, it will rarely persist. It is distinguished from other prairie milkweeds by its narrow leaves and very thin seedpod. The seeds can be purified by the procedure described for *A. purpurascens*. The seeds require 2 months of cool-moist stratification or no treatment for germination [11]. Fall planting also is satisfactory. Its propagation is recommended only if plant diversity is desired.

2.16 *Aster* see *Symphyotrichum*

2.17 **Astragalus canadensis* v. *canadensis* (Fabaceae Family), Canadian Milk Vetch

Astragalus canadensis v. *canadensis*, Canadian milk vetch, (height, 0.3–1.4 m) blooms from late June to mid July with many greenish-cream to white flowerets in a dense spike. See illustration on page 104. Canadian milk vetch is found throughout the U.S., except the upper Northeast and the West. In Canada it is found from Quebec west to British Columbia and north to the Northwest Territory. Two other varieties are located in the West. The pinnately compound leaves offer some variety in appearance. Canadian milk vetch has a fairly high Coefficient of Conservatism of 7, and its eco-preference is full sun or partial shade in wet-mesic to dry-mesic soils. In mid August to early September clusters of black pods can be collected. Under certain circumstances of weather, two crops of pods can occur, the latter requiring collection in late October. After drying, the thick-walled pods (about 1–2 cm long) release several tiny bean-like seeds (2 mm long or less). More complete seed recovery is obtained after the pods are crushed. Seeds can be easily separated from most of the crushed pods by sieving through ordinary window screen. This plant is a typical legume, which does best with the scarification of the seed coat followed by 10 days of cold-moist stratification [11]. Before sowing, an inoculation with specific nitrifying bacteria is recommended [11]. Fall or early spring sowing without seed treatment also afforded a rather low percentage germination, in which case additional seeds germinated after the second year.

2.18 *Aureolaria grandiflora* v. *pulchra, Gerardia grandiflora* (Scrophulariaceae Family), False Foxglove

Aureolaria grandiflora v. *pulchra*, false foxglove, (height, 0.4–1.2 m) is not a true prairie plant as it is found in oak savannas, but it seems to be generally associated with adjacent high-quality prairies (illustration on page 104). The variety *pulchra* is found in six states of the upper Midwest from Indiana west to Minnesota and south to Missouri. Three other varieties are located in the states of the southern Midwest. False foxglove is partially a root-parasite on oaks. Although it has a Coefficient of Conservatism of only 6, it does associate with oaks. A large yellow, foxglove-like flower, occurring about mid-August, makes this plant special in savanna settings. The relatively large seed capsules (about 1.5 cm long) can be collected from mid-October to early November, from which seeds (about 2 mm long) can be recovered by crushing the capsule. During the fall, seeds can be sown on top of the soil after shallow scraping of the soil within the outer dripline of an oak tree. This method of planting gives results, but the germination rate is somewhat low. According to Deno [12], seeds sown outdoors in January germinated in April. Its soil requirement is simply where oak trees grow, that is, from mesic to dry-mesic soils.

2.19 **Baptisia alba* v. *macrophylla, Baptisia leucantha/lactea* (Fabaceae Family), White Wild Indigo

Baptisia alba v. *macrophylla*, white wild indigo (height, including the flowering spike, 1–2 m), is a spectacular plant that emerges only after all danger of frost is over, usually in June. See illustration on page 104. The initial emergence is reminiscent of asparagus, but shortly after, in late June, it opens into a leafy canopy surmounted by a spike of multiple white flowers. This attractive legume is a must in all prairie restoration projects, and will grow in full sun or savannas on wet-mesic to dry soils. The *macrophylla* variety is found from New York west to Minnesota and Nebraska and south of this northern occurrence to the border, with the exception of the southeastern and middle Atlantic states. White wild indigo has a Coefficient of Conservatism of 6, and it is not difficult to propagate even with competition from tallgrass. This plant has deeper taproots than tallgrass, probably accounting for its success in competitive prairie ecosystems. Interestingly, it will grow in shallow soil underlain with shale. Clusters of black, thick-walled pods (2.5–4 cm long) containing several seeds can be harvested from late August to late September. The pods can be recovered either by clipping the pod stem or by simply hand-stripping the pods from the stem. Because the seedpods are often contaminated with weevils, the bean-like seeds (3–5 mm long) should be obtained as soon as possible by crushing the pod. Although the seeds can be somewhat separated from the crushed pods by use of quarter-inch hardware-cloth, a sticky resinous substance associated with the seed makes efficient recovery difficult. For this reason, it is hardly worthwhile to separate the seeds from the crushed pods. Because of weevil infestation, one should add powdered Sevin insecticide before seed storage; thus, chemical-resistant gloves and a facemask should be worn while sowing. Alternatively, the seed can be refrigerated before or after processing. Curiously, weevils have not yet been found in seeds grown in the East. The recommended seed treatment is scarification followed by a 10-day cold-moist stratification and inoculation with specific nitrifying bacteria [11]; however, with this species nitrifying bacteria have been found to be unnecessary. A planting in late fall without any seed treatment is both facile and gives good results (about 55%). About 35% of seeds germinated during the second springtime, and 10% of seeds will germinate during the third season. Late emergence of white wild indigo is convenient when growing the plant as a seed producer. Weeds can be eliminated from beds by glyphosate herbicide before early May.

One other variety, *Baptisia alba* v. *alba*, occurs in the Southeast south from Virginia.

2.20 **Baptisia australis* v. *australis* (Fabaceae Family), Blue Wild Indigo

Baptisia australis v. *australis*, blue wild indigo (height, 1.2–1.6 m), is found from New Hampshire west to Wisconsin and all states south of this northern limit, except Mississippi and Florida. See illustration on page 104. Interestingly, two populations of *B. australis* reportedly existed in high-quality prairies near Chicago [16], certainly indicating that this variety is native in northern Illinois. A smaller variety, *B. australis* v. *minor*, is found from Kentucky and south to Texas and Arkansas. A third variety, *aberrans*, is found only in Tennessee, North Carolina, and Georgia. The preferred habitat is alluvial savannas, but it thrives on mesic to dry-mesic prairies. Because of the attractive aspect of the plant and blue flowers, blue wild indigo is recommended for restorations. Blue wild indigo blooms in late May when most prairie forbs are not yet in flower, which gives added incentive to include this forb in prairie plantings. The optimum seed germination is achieved at 70°F (90–95% in 1–10 weeks) using either fresh seed or seed dry stored for 6 months at 40 or 70°F [12]. Seed scarification of fresh or 6-month-stored seed lowered the percent germination. However, seeds dry stored for 1 year required a seed coat puncture (scarification) to germinate 100% in the second day at 70°F in the dark. If not punctured, 10% germination occurred on the second day. In my hands, fall planting seeds of unknown age resulted in a relatively small percentage of spring germination.

2.21 **Baptisia bracteata* v. *leucophaea*, *Baptisia leucophaea* (Fabaceae Family), Cream Wild Indigo

Baptisia bracteata v. *leucophaea*, cream wild indigo (height, 0.2–0.5 m), affords spectacular cream-colored flower clusters in mid-May, which is rather early for most prairie forbs (illustration on page 104). It can be a special sight in the presence of blue and red colors of lupine and Indian paintbrush blooming at the same time. This special sight is accentuated after an early spring prairie fire clears plant debris from the site. In this author's opinion, cream wild indigo is one of the most attractive of the prairie forbs. Except for its presence in New Jersey and Massachusetts, cream wild indigo is localized in the Midwest. In the north it is found from Michigan west to Minnesota and South Dakota, and its southern limit is from Mississippi west to Texas. Restoration is somewhat difficult, as indicated by its Coefficient of Conservatism of 9. However, propagation in weeded beds or row crops was relatively easy. Cream wild indigo does well in full sun or savannas, but the soil type is critical. Hillsides and mesic to dry soil are best. Lack of competition from other plants, such as poor soil, aids restoration efforts. Low-lying clusters of dark-brown relatively thin-walled pods can be harvested in early August or whenever the pods become brown. Because rodents scavenge for the seeds, one should not wait too long to harvest. The seedpods are somewhat larger than described above for *B. alba* v. *macrophylla*, but otherwise the seed collection procedure is the same, furnishing bean-like seeds almost completely resembling *B. alba* v. *macrophylla*, with the exception that the sticky substance surrounding the seeds is absent in *B. bracteata* v. *leucophaea*. Thus, sieving through quarter-inch hardware cloth is easy with cream wild indigo. Like the seeds of *B. alba* v. *macrophylla,* the seeds of cream wild indigo are often infested with weevil, requiring treatment with either insecticide or refrigeration. Seed treatment for germination should follow the same protocol described above for *B. alba* v. *macrophylla.* When seeds of this species were spring- or fall planted without seed treatment, poor germination occurred the next growing season. However, the seeds germinated remarkably well in the spring of the following year. This author prefers planting/sowing in the late fall without any prior treatment. Deno [12] reported germination in April from sowing dry stored seeds outdoors in December. Prairie Moon Nursery recommends seed scarification, 10 days of cold-moist stratification, and inoculation with nitrifying microorganisms [11]. In my opinion, inoculation with nitrifying microorganisms is unnecessary.

2.22 **Brickellia eupatorioides* v. *eupatorioides*, *Kuhnia eupatorioides* (Asteraceae Family), False Boneset

Brickellia eupatorioides v. *eupatorioides*, false boneset (height, 0.3–1 m), is a rather inconspicuous plant, even when in flower (see page 104 for illustration). This plant can be added to restoration projects if diversity is desired; otherwise, it is not remarkable. Plants emerge later than most prairie plants. The flowers are clusters with small creamy-white flowers. False boneset favors mesic to dry soil in full sun or partial shade. Although it can grow on mesic soils, it prefers drier conditions. The typical variety *eupatoroides* extends its northern range from New Jersey westward past the Great Lakes to Illinois, Missouri, and Texas. It is found in all states south and east of its northern range. Of the five additional varieties, three are absent in the tallgrass area. Two others, v. *corymbulosa* and v. *texana*, are found in the Midwest and eastern Great Plains. False boneset's Coefficient of Conservatism is 6. Seed-heads can be clipped from late August to mid October when the fluffy dull-white pappus appears. After a short drying period, the seeds plus pappus (1 cm long) can be readily thrashed loose or recovered by scraping across quarter-inch hardware cloth. The seed itself is dark brown (5 mm long and less than 1 mm wide). Seeds need no treatment for germination [11], but they respond well to sowing in the late fall.

2.23 *Cacallia* see *Arnoglossum*

2.24 **Camassia scilloides* (Liliaceae Family), Wild Hyacinth

Camassia scilloides, wild hyacinth (height, 0.3–0.6 m), is one of the earliest flowers to bloom on the prairie (mid-May). See page 104 for illustration. Often deer deprived of forage over the winter will decimate both the flower and leaves. This plant prefers wet-mesic to mesic soils in either full sun or partial shade. It ranges west along a northern limit from Pennsylvania, Ohio, and Ontario to Wisconsin, Iowa, and Kansas. It is also found south of this northern limit, except Florida. A Coefficient of Conservatism of 7 indicates a fair intolerance of disturbance. Multiple capsules or fruiting bodies form on the flower stem, which ripen in mid-June. When the upright capsule opens to reveal the shiny black seeds within, the capsules can be stripped from the stem, or the entire flower head can be clipped. Be careful in this step as loose seeds are fairly easily dropped. The black shiny seeds are oblong spheres about 2–3 mm long. In my opinion, the seeds are best fall planted. Alternatively, a 2-month cool-moist stratification is required for germination [11]; however, this author has observed wild hyacinth seeds germinating in moist sand while being refrigerated. As this is a cool season plant, perhaps it is not surprising to see cold germination. If the cool-moist stratification method is chosen, one should periodically check the seeds.

2.25 **Carex brevior* (Cyperaceae Family), Plains Oval Sedge

Carex brevior, plains oval sedge (height, up to 1 m), is only one of more than 80 species of *Carex* found in either savannas or prairies, of which about two dozen are common (see common *Carex* species in Appendix 1 marked with an asterisk). Obviously, this genus is underrepresented in this treatment. With a few exceptions, those species selected for description are some of the most commonly found. The *Carex* family is characterized by a three-angled solid stem, which supports the inflorescence/seed-head (culm). Many *Carex* species are somewhat indistinguishable, and sometimes identification relies on the overall aspect, as well as the structural appearance of the seed-head and its components [see Mohlenbrock [17]]. Principal features of seed-head components include the perigynia (a sack-like covering enclosing the achene), the achene (a one-seeded dry fruit), and the pistillate scale (the chaffy part on the outermost side of the "seed"). *Carex brevior*, plains oval sedge, is fairly common in dry soil habitats, such as sands or railroad beds, but it will tolerate wet-mesic to dry soils. See page 104 for illustration. This *Carex* has a preference for full sunlight to partial shade. It ranges over southern Canadian provinces and all the U.S., except for the Southeast and much of the far West. A Coefficient of Conservatism of 4 indicates a medium tolerance of disturbance. Seed-heads can be collected from mid- to late June. After drying, the seeds can be recovered by thrashing or scraping over quarter-inch hardware cloth. As for most *Carex* species, the recommended seed treatment for germination is 2 months cool-moist stratification and sowing on top of the soil [11]. It is recommended that seeds lying on the surface for germination should not be permitted to dry. Various methods can be used to prevent soil drying, such as a sheet, burlap, screen, or a sparse scattering of dried grasses or straw [11]. If the latter is chosen, be careful that weed seeds are not introduced with the dried grass/straw. I have found that fall sowing on glyphosate-treated grass is very successful, with plants appearing the second growing season. Other commonly found *Carex* species possessing an inflorescence/seed-head similar to *C. brevior* are *C. cristatella*, *C. festucacea*, and *C. molesta*, all of which differ from *C. brevior* in their preference for moist habitats.

2.26 **Carex davisii* (Cyperaceae Family), Davis's Sedge or Awned Graceful Sedge

Carex davisii, Davis's sedge or awned graceful sedge (height, up to 1 m), is found in wet-mesic to mesic situations in open sunlight to shaded habitat (illustration on page 104). It is relatively common and easy to establish, as indicated by its Coefficient of Conservatism of 3. This sedge is located in the north from Massachusetts west to Ontario, and North Dakota, and from this northern limit it is found south to Texas and Arkansas. It is absent in the Southeast from Virginia to Louisiana. Seed-heads can be harvested from middle to late June. For successful germination seeds can be cool-moist stratified for 2 months or fall planted. Seed recovery and sowing follows the procedure recommended for *C. brevior* above.

2.27 **Carex gravida* v. *lunelliana* (Cyperaceae Family), Heavy Sedge or Long-Awned Bracted Sedge

Carex gravida v. *lunelliana*, heavy sedge or long-awned bracted sedge (height, up to 1 m), is fairly common in the Midwest, inhabiting mesic to dry prairies or savannas (illustration on page 105). Its preference for drier conditions distinguishes it from the majority of wet-loving *Carex* species. The variety *lunelliana* is absent in the Northeast north of Maryland, and it is not found west of the Rocky Mountains. The variety *gravida*, ranging somewhat more north, is identified by a more oblong perigynium as opposed to a more spherical shape observed with v. *lunelliana* [17]. A Coefficient of Conservatism of 4 indicates that this species is fairly tolerant of disturbance. Seed-heads can be collected from early to late June. After drying it is easy to dislodge the parts of the seed-head (perigynia plus achene) simply by scraping the seed-heads over quarter-inch hardware cloth. The seeds are too large to pass through an ordinary window screen, but a sieve with somewhat larger holes is required for further seed purification. Seeds require a 2-month cool-moist stratification for germination followed by sowing on top of the soil. Fall planting was somewhat successful. Copious volunteer plants surrounding the nursery beds from natural seed drop are a good indication that the seeds can be sown immediately after June collection. Other common *Carex* species somewhat resembling the inflorescence/seed-head of *C. gravida* are *C. cephalophora*, *C. mesochorea*, and *C. muhlenbergii*, all of which are found in similar habitats.

2.28 **Carex shortiana* (Cyperaceae Family), Short's Sedge

At maturity the green seed-head of *Carex shortiana*, Short's sedge (height, up to 0.9 m), changes to dark brown, giving a readily identifiable characteristic (illustration on page 105). As seen in the illustration, the seed-head also resembles the tightly packed segments of a pinecone. This species is found in wet-mesic to mesic areas in full sun to shaded situations. It is found in the states of the Middle Atlantic from New York south to Virginia and in the Midwest from its northern range in Ohio and Ontario west to Iowa and in the south from Tennessee to Oklahoma. A Coefficient of Conservatism ranked at 4 indicates a medium ease of restoration and tolerance to disturbance. Seed-heads can be collected after changing to a brown color from late June to mid-July. After drying, the perigynia plus achenes (seeds) are readily dislodged with little effort. For germination, seeds require a 2-month cool-moist stratification, after which the seeds should be sowed on top of the soil [11]. Soil drying should be avoided (see for procedures Sect. 2.25 above).

2.29 **Carex vulpinoidea* v. *vulpinoidea* (Cyperaceae Family), Fox Sedge

Carex vulpinoidea v. *vulpinoidea*, fox sedge (height, 1–1.2 m), one of the most common and widespread *Carex* species in North America, is found all over the U.S. and southern provinces of Canada, except Utah. See illustration on page 105. One other variety is restricted to Indiana, Michigan, Minnesota, and Ontario. It is very easy to restore on most soils from wet to dry-mesic, particularly if a wet-mesic habitat in bottomlands is provided. Although this *Carex* is adaptive by its survival in fairly drastic conditions of dryness, under these conditions it will decrease in density. A Coefficient of Conservatism of 3 indicates its ease of restoration. The inflorescence first appears in late May. This *Carex* can be identified by the aspect of the greenish or straw-colored seed-head caused by the color of the perigynia (the covering surrounding the seed or achene), and its relatively thin leaves (2–6 mm wide). Seed-heads can be collected from late June to mid-July. Seeds (perigynia plus achenes) can be easily removed from the dried seed-heads by thrashing or scraping on quarter-inch hardware cloth. A 2-month cool-moist stratification of seeds followed by sowing on top of the soil is recommended [11]. Or, sowing seeds in the fall is satisfactory.

A related species, **Carex annectens*, yellow fox sedge (not shown), has a yellow or golden-brown perigynia (layer covering the achene or seed); hence, the seed-head has a golden-brown color. Moisture requirements and seed treatment of *C. annectens* are essentially the same as for *C. vulpinoidea* [11]. Other common *Carex* species somewhat resembling the inflorescence/seed-head of *C. vulpinoidea* are *C. conjuncta* and *C. stipata*, both of which prefer growing in wet soils.

2.30 *Cassia* see *Chamaecrista* and *Senna*

2.31 **Castilleja coccinea* (Scrophulariaceae Family), Indian Paintbrush

In early to mid-May, *Castilleja coccinea*, Indian paintbrush (height, 0.2–0.5 m), is a special sight on prairies with its scarlet flowers (illustration on page 105). With a relatively high Coefficient of Conservatism of 8, it is a scarce item on most prairies, mainly because of its requirements for survival. According to some authorities [13–15] it is an annual or biennial plant, but my observations indicate that it may be a perennial. It is a parasite on the roots of certain grasses, such as *Schizachyrium scoparium*, little bluestem, or *Koeleria macanthra*, June grass [11]. However, I have noted its growth in the apparent absence of grass, such as in a dense stand of *Eryngium yuccifolium*, an indication that its requirements need more study. Its preferred ecosystem is wet-mesic to mesic peaty or sandy/gravelly soil. Indian paintbrush occasionally can be observed on dry-mesic sand/gravel hillsides or on mesic hillsides with ordinary soil, provided that its hosts are present. Although Indian paintbrush is scarce over most of its range, it may be found from Maine west to Minnesota and Kansas and south of its northern limit. It also exists in mid-southern provinces of Canada. The seed appears like sawdust, and can be collected when the flower head senesces to a brownish-tan color in late June. It is recommended that seeds should be sowed at the base of the host grass after a 2-month cool-moist stratification [11]. However, it is satisfactory to simply scatter the seed in areas of the host grass or in a mature prairie ecosystem shortly after seed collection.

Another species of *Castilleja* is worthy of mention, but very difficult to grow. With a Coefficient of Conservatism of 10, *Castilleja sessiliflora*, downy painted cup (height, 15–35 cm, not shown), is rare enough to be listed as endangered in Illinois. It flowers in mid-May. This author has only seen the latter species in the sandy hills of Nachusa Grasslands in Illinois, a project of The Nature Conservancy. Even there, it is relatively inconspicuous in appearance and abundance, which indicates that success with downy painted cup would be likely in full sun on dry-mesic to dry sandy soils in association with its host grass, such as little bluestem. In contrast to *C. coccinea*, *C. sessiliflora* is a perennial plant. Except for Wisconsin and Illinois, its range is west of the Mississippi River.

2.32 **Ceanothus americanus* (Rhamnaceae Family), New Jersey Tea

Ceanothus americanus, New Jersey tea (height, 0.2–1 m), is one of the few woody shrubs found on the prairie, which regenerates from the root after a prairie burn (illustration on page 105). The leaves of New Jersey tea served as a substitute for tea during the American Revolution. New Jersey tea is found from Maine west to Minnesota and Nebraska, and from this northern limit extends south to the border. It is also found in Quebec and Ontario. Plants of the *Ceanothus* genus are one of the few non-legume species that host nitrifying microorganisms in their roots [18]. It prefers mesic to dry-mesic hillsides in full sun or partial shade. It is often found on gravelly soil or black prairie silt-loam, leading one to suspect that loose friable soil is preferred but not absolutely required. A Coefficient of Conservatism of 8 indicates that this plant is not amenable to much disturbance. A dark gray fruit (husk) about 4–5 mm in diameter contains three oblong dark-brown seeds (2 mm long); this husk must be removed for efficient germination. This author does not recommend the boiling water procedure for stimulating germination [11]. In my hands the use of boiling water was a total failure. According to experiments by Deno [12], hot water between 150 and 180°F is fatal to seeds, and it is more effective to physically produce a hole in the seed coat. Alternatively, the seeds can be scarified followed by 70 days cold-moist stratification [11]. According to another source [3], cuttings can be placed in moist sand. Yet another alternative [12] is a 40°F moist treatment for 3 months (1% germination) followed by 70°F (28% germination in 1–21 days). Therefore, fall plantings should be satisfactory. Herbivores, probably deer, often decimate the leafy branches and inflorescences.

2.33 **Chamaecrista fasciculata* v. *fasciculata*, *Cassia fasciculata* (Fabaceae Family), Partridge Pea

Chamaecrista fasciculata, partridge pea (height, 0.25–0.75 m), is one of a minority of annual plants found in prairies (illustration on page 105). With a Coefficient of Conservatism of 1, partridge pea is an opportunistic plant that forms huge colonies in the absence of severe competition from other plants. In flower these colonies give a profusion of bright yellow color. It is usually found in the full sun in wet-mesic to dry-mesic soil, usually in disturbed or poor soils. It is surprisingly sensitive to drought. Partridge pea is located from Massachusetts west to Minnesota and South Dakota and south of this northern range to the U.S. border, including New Mexico. One other variety is restricted to Maryland and Virginia. Flowering of partridge pea occurs in early August. The brown seedpods can be collected in mid-September by stripping them off the plant. Pods should be placed in a paper bag to dry (preferably in full sunlight) with the bag top folded over or stapled shut. Drying causes the pods to split open and twist, which sends the seeds flying for some distance. The shiny black seeds are somewhat flat and rectangular, measuring about 3 mm long. The recommended seed treatment is typical of legumes, that is, scarification of the seed coat, 10 days cold-moist stratification, followed by inoculation with specific nitrifying bacteria [11]. However, I have found that it is not necessary to inoculate with nitrifying bacteria. Fall planting in an ecosystem described above is satisfactory for success, but without scarification usually more than one growing season is required for germination of the majority of seeds.

Compared to partridge pea, a very similar plant, **Chamaecrista/Cassia nictitans*, wild sensitive-plant (see herbarium and seeds/seed-head on Springer internet site), is found in the East and Midwest, and is distinguished from partridge pea mainly by its much smaller flower (4–8 mm broad). Very similar seed-pods ripen about the same time as partridge pea. Pod splitting causing seed shatter is virtually identical to that encountered with partridge pea. Fall planting gives good results; thus, seed coat scarification seems to be less important with this species.

2.34 **Cirsium discolor* (Asteraceae Family), Field or Pasture Thistle

Cirsium discolor, field thistle (height, 1–3 m), is mentioned simply because it is often an adventive native found on most prairies, especially in recently restored sites. See illustration on page 105. Field thistle is found from Maine west to Minnesota and south of this northern limit, except for Florida. It is also found in the Dakotas and southern Canada from the east to mid-provinces. A Coefficient of Conservatism of 3 is indicative of its ability to tolerate a fair amount of disturbance. Field thistle prefers to grow in wet-mesic to dry-mesic soils and full sun or partial shade. Some tolerance of this plant is recommended, but active restoration is not desirable or necessary.

For germination, recommended seed treatment is 2-month cold-moist stratification [11]. One should be vigilant for aggressive non-native thistles, which should be eradicated before they spread. Some of these aggressive Eurasian species are *Cirsium arvense* (Canadian thistle), *Carduus nutans* (musk thistle), and *Cirsium vulgare* (bull thistle). *Cirsium discolor* in overall appearance seems similar to the European import *C. vulgare*, but *C. discolor* is distinguished by the whitish underside of the leaves. In addition, the upper stem of *C. vulgare* is covered with prickly broad wings, whereas the upper stem of *C. discolor* is largely without prickly wings. One should be careful to preserve the very rare *Cirsium hillii*, Hill's thistle, which is often mistaken for other thistles. This relatively small plant, described in Chap. 3, is found on old established hill prairies.

2.35 **Comandra umbellata* ssp. *umbellata, Comandra richardsiana* (Santalaceae Family), Bastard or False Toadflax

Comandra umbellata ssp. *umbellata*, bastard toadflax (height, 0.15–0.4 m) is generally associated with old established prairies or savannas; thus, the assigned Coefficient of Conservatism of 6 seems to be far too low. See illustration on page 105. It ranges in the north from Maine west to North Dakota, including much of Canada, and in the south from Georgia west to Oklahoma. Although one authority [14] considers *C. umbellata* to be a separate species from *Comandra richardsiana*, most categorize the two as the same (e.g., [13, 15]). There are two other subspecies, both ranging further west. Bastard toadflax is a general parasite on the roots of other plants, not necessarily specific grasses as required by *Castilleja* species. Its preferred habitat is mesic to dry fairly sterile sandy/gravelly soil in full sun or partial shade. It does survive on brown forest silt-loam soil as well. White flowers appear in early to mid-May. The seed is contained in a spherical green fruit 4–6 mm in diameter, but the fruit tends to be very sparse. Seed/fruit is ready for harvest in July. The plant mainly propagates by spreading root rhizomes. Although it is not generally advisable to encourage transplanting diggings of prairie plants, bastard toadflax responds well to early season transplantation. This author transplanted a sample from the path of a freeway development, and subsequently this single transplant spread into a large patch that served as a source of other restorations. If the fruit is used, it is recommended that each fruit be pressed into the soil in an established plant community, where bastard toadflax can find opportunity to parasitize other plants. Poorer-soil less-dense plant communities are best for propagation from fruit.

2.36 *Coreopsis palmata* (Asteraceae Family), Prairie Coreopsis

Coreopsis palmata, prairie coreopsis (height, 0.3–0.9 m), is found on mesic to dry prairies but it suffers under drought conditions. See illustration on page 106. It is a Midwestern species ranging in the north from Michigan west to Minnesota and

South Dakota. In the south its limit is Alabama west to Oklahoma. Prairie coreopsis has a fairly high Coefficient of Conservatism of 6. Yellow flowers appear in late June. Because the plants spread by rhizomes, they are usually found in dense colonies. The almost black senescent seed-heads can be clipped around mid-September to early October. The seed-heads are very difficult to break open to collect the seeds and usually require a tedious shelling of the seeds by hand from each seed-head. An alternative involves crushing with the top end of a sledgehammer. If one can risk predation by birds, collection of seed-heads in mid-November to January affords a much easier recovery of the seeds. The seeds are dark brown, flat, and oblong in shape, 2 by 6 mm. For germination the seeds require 2 months of cool-moist stratification [11]. Alternatively, seeds can be sown in the fall.

2.37 **Coreopsis tripteris* (Asteraceae Family), Tall Coreopsis

Coreopsis tripteris, tall coreopsis (height, 1–3 m), is found on open prairies, but more often in partial shade (see illustration of plant on page 106, including an inset showing the multiply branched flower). Its northern limit is Massachusetts west to Quebec, Ontario, Wisconsin, Iowa, and Kansas. The southern limit in the U.S. is Florida west to Texas. Apparently it is absent in New York, New Jersey, and Delaware. Because this plant has interesting leaf structure, it is in my opinion more visually appealing before flowering. Its preference for soil is from wet-mesic to dry-mesic. Its Coefficient of Conservatism of 4 indicates a relative ease of restoration. Initially, it does well on open prairies, but decreases somewhat as the competition increases. Because row crops persist for some time, tall coreopsis most likely has more than a decade of life. The flower blooms in early September, and the seed-heads can be clipped when brown from early October to late October. Goldfinch feed heavily on the seed, so collection should not be delayed. The seeds can be removed from the seed-heads by crushing underfoot. Alternatively, seeds are easily removed from seed-heads by scraping and sieving through quarter-inch hardware cloth. The gray-brown seeds are flat oblong in shape (2 by 6 mm), and they require 2 months of cool-moist stratification [11] or fall planting.

2.38 *Dalea candida* v. *candida*, *Petalostemum candidum* (Fabaceae Family), White Prairie Clover

As indicated by its Coefficient of Conservatism of 9, *Dalea candida* v. *candida*, white prairie clover (height, 0.3–0.7 m), is difficult to establish in new restorations. See illustration on page 106. It is found in its northern range from Ontario west to Alberta. In the U.S. it is found from Indiana west to North Dakota. The southern range extends from South Carolina west to Texas. It is absent from most of the eastern coastal states from Maine south to Virginia, including Florida. One other variety, *oligophylla*, has a somewhat more western occurrence with its easternmost range in Wisconsin. White prairie clover is a difficult species to restore. The possibilities are better when restoration is attempted in an established prairie where the competition is not too great in soils ranging from mesic to dry (good drainage). This plant also responds well to cultivation with regular weeding. Because the seedlings are small, careful weeding is required, or preferably seed should be sowed on weed-free soil. Susceptibility to grazing by herbivores has been noted. The plant flowers in early to mid-July. The seed-head becomes senescent brown in early September to mid-October, and the densely packed bracts plus pods (each about 4 mm long) can be collected when the stem of the seed-head becomes senescent brown just under the seed cluster. At this time the seed offers minimal resistance to hand stripping. The bract plus pod (hull) contains an individual seed (1.5–2.5 mm long). Although stratification is not needed, hull removal is recommended [11], but an overly vigorous hull removal by friction can damage the seed. The hull can be removed by rubbing over a metal screen backed by a slightly recessed surface (to prevent destroying the screen but permitting the seed to pass through) until the dehulled seed pass through the screen. It is this author's opinion that of all the legumes, *Dalea* species benefit the most from inoculation with specific *Rhizobium* microbes before sowing, which can be obtained from prairie nurseries, such as Prairie Moon Nursery. In my hands fall planting is not very successful; thus, a spring sowing is better. It is recommended that extra effort should be lavished on this conservative species by growing potted plants by the procedure outlined for *A. canescens* above, especially with limited quantities of available seeds. Plants established in this way will propagate others in due time.

2.39 *Dalea foliosa, Petalostemum foliosa/foliosum* (Fabaceae Family), Leafy Prairie Clover

The truly rare *Dalea foliosa*, leafy prairie clover (height, 0.3–0.6 m), is federally listed as endangered, and consequently has the highest Coefficient of Conservatism of 10 (illustration on page 106). It is only native to Wisconsin, Illinois, Tennessee, and Alabama. Although there are reasons for restricting the sale of leafy prairie clover by nurseries, it would seem to be a certain recipe for its extinction. In northern Illinois, *D. foliosa* is found in mesic to dry-mesic dolomite hill prairies along rivers. This author has had good success in cultivating the plant in mesic soil from seed originally obtained from a nursery before restriction by law. Periodic weeding is necessary to reduce competition. The plant also has been grown on a dry-mesic south-facing hillside where there is reasonably low competition. During very wet conditions a damping-off has been noted, possibly a susceptibility to fungus. Leafy prairie clover does not seem to have a lifetime of more than 5 years. Wildlife feeds on the foliage, and perhaps rodents feed on the roots over winter. Flowering occurs in mid-July, and tightly packed bracts plus pods strip off the seed-head with little resistance in late August to mid October. If resistance is encountered, let seed-heads ripen further (the stem of the seed-head should be senescent brown). The bract plus seedpod (hull) is about 4 mm long, and the tan-to-brown seed (about 2 mm long) is contained within. For germination, removal of the hulls or pods is recommended by the method described for *Dalea candida* or *Amorpha canescens*. Seeds should be inoculated with nitrifying bacteria specific for *Dalea* species before sowing in the spring. This author has had some success by a late fall sowing of the seeds with the intact hull. By this method, expect some germination in the spring and summer, and additional seeds may germinate the second year. Norman Deno (personal communication) found that germination occurred within 2 days after removing the hull followed by careful scarification of the tiny seeds by dragging them over sandpaper.

2.40 **Dalea purpurea* v. *purpurea, Petalostemum purpureum* (Fabaceae Family), Purple Prairie Clover

Dalea purpurea v. *purpurea*, purple prairie clover (height, 0.3–0.9 m), affords attractive bright purple flowers in mid-July (see page 106). It is found in mesic to dry prairies and savannas on well-drained soil. The northern extent of the typical variety *purpurea* is from New York and Ontario west to Alberta and Montana. From this northern area all states directly south and east are included in its range, except most of the eastern states, as well as Idaho and Utah. New York and Georgia are the only eastern states where v. *purpurea* has been reported. One other variety is found in a few states of the Mid-South and Great Plains. The Coefficient of Conservatism for purple prairie clover is a fairly high 8. Tightly packed bracts plus the seed pods can be stripped by hand from the seed-head only when it offers little resistance to removal, about late August to early October (the stem just below the seed cluster should be senescent brown). These grayish bracts plus pods (4–5 mm long) have a distinctive odor similar to lemon peel when freshly stripped. The pod contained within the bract in turn contains a brown spherical seed (1.5–2 mm long). It is recommended to remove the hull material, but care should be taken to not destroy the seeds. See Sect. 2.38 or 2.4 for procedures to remove the hull. Inoculation with nitrifying bacteria specific for *Dalea* is recommended before sowing; no other treatment is required [11]. In my hands a late fall planting is ineffectual and/or variable. Because the seedling is easily overwhelmed by lush weed growth, purple prairie clover is difficult to establish. It is preferable to select a site with little weed competition, such as an established hillside prairie with patches of bare earth visible, or hope for a dry growing season. Sod removal, which eliminates many weed seeds, is an alternative that has been occasionally used by some growers, such as Peter Schramm of Galesburg, Illinois. Although in the East I have successfully grown this plant in weeded beds, establishment in the field has been very difficult. Thus, I have resorted to growing purple prairie clover in pots before placement in the field (see Sect. 2.38 or 2.4 for procedures).

2.41 **Desmanthus illinoensis* (Fabaceae Family), Illinois Bundleflower or Prairie Mimosa

Desmanthus illinoensis, Illinois bundleflower (height, 0.3–2 m), has lacey bipinnate leaves typical of the mimosa family (illustration on page 106). To some extent the leaves are somewhat sensitive to touch or rain, causing the leaflets to slowly fold. Illinois bundleflower has a much larger range than just Illinois. Its northern range is Pennsylvania west to Illinois, Wisconsin, Minnesota, North Dakota, Colorado, Utah, and Nevada. From the northern range it is found in all states south and east, except Arizona, New Jersey, Delaware, and West Virginia. With a Coefficient of Conservatism of only 4, Illinois bundleflower is easily restored in mesic to dry-mesic prairies. This perennial is not exceptionally long lived. It flowers in mid-July, and the seed-heads are ready for harvest from mid-September to mid-October; however, the seed-heads tend to remain intact for some time later. The seed-heads can be clipped from the plant or stripped off as clusters of pods by hand. The seeds are contained in numerous curved legume pods (1.5–2.5 cm long by about 5 mm wide) in bundled globose heads. Two to six shiny brown oblong seeds (3 mm long) are contained within the pods. Some vigorous scouring or scraping action is required to break up the pods, which resist release of the seeds. Scarification of the seed coat followed by inoculation with specific nitrifying bacteria (no cool-moist stratification) is recommended for germination [11]. Actually, a low to fair percentage of success in germination is obtained simply with no treatment at all after removal of the pods. Fall planting is also an option. It has been my experience that seed germination can be variable. The fungal-produced germination hormone, gibberellic acid-3, aids germination at 70°F [12]. Gibberellic acid-3 with instructions for use can be purchased from certain nurseries. Wes Jackson of the Natural Land Institute in Kansas considers this plant to have promise as a perennial agricultural seed crop. To be viable as a crop, considerable genetic manipulation would be necessary to increase the size of the seed and ease of harvest.

2.42 **Desmodium canadense* (Fabaceae Family), Showy Tick Trefoil

Desmodium canadense, showy tick trefoil (height, 0.6–2 m), has the showiest flowers of the *Desmodium* genus (illustration on page 106). The name tick trefoil undoubtedly derives from the tick-like appearance of the pod segments that stick to everything, much like Velcro. Coincidentally, the inspiration for invention of Velcro was from study of the mechanism of "stick-tight" seeds. The northern range is New Brunswick west to Manitoba and North Dakota. It occurs south of the northern boundary, except that it is absent in nine states of the Southeast. With a Coefficient of Conservatism of 5, showy tick trefoil is very easily restored on wet-mesic to dry-mesic soils in full sun or partial shade. The plant seems to resist drought very well, but under these conditions seed production is sparse. Also, Japanese beetles feed on the flowers, which greatly prevents seed set. Flowering occurs in early to mid-August with the brown seed pods easily hand-stripped from the plant in early to mid-September. The seed-pods are jointed in three to five segments, easily breaking into individual almost triangular segments 4–5 mm long that contain one seed. The recommended seed treatment is pod removal followed by inoculation with specific nitrifying bacteria [11]. However, this species is so easy to restore that this author simply sows the pod segments containing the seeds in late fall with no other treatment. The only problem is the tendency for the pod segments to cling together. To alleviate this problem, mix thoroughly with either sand or other seeds until the seeds are reasonably well separated. Deno [12] found 100% germination at 70°F within 2 days only when the seed coat was punctured; otherwise, no germination was observed. Evidently, fall planting permits water to penetrate the seed coat over winter.

Desmodium illinoense, Illinois tick trefoil (see herbarium on Springer internet site; height, 0.6–1 m), is similar to *D. canadense* with slightly different ecosystem requirements of full sun and mesic to dry soils. Illinois tick trefoil's habitat is restricted to the Midwest. Compared to *D. canadense*, the white to pinkish-white flowers of *D. illinoense* are not as showy and the pod segments are more rounded (see seed-head and seed-pods on Springer internet site). Restoration of Illinois tick trefoil is the same as showy tick trefoil.

2.43 **Dichanthelium oligosanthes* v. *scribnerianum*, *Panicum oligosanthes* v. *scribnerianum* (Poaceae Family), Scribner's Panic Grass

Dichanthelium oligosanthes, panic grass (height, 0.1–0.6 m), is found as two varieties. The most common of these is *D. oligosanthes* v. *scribnerianum,* Scribner's panic grass. See illustration on page 106. Scribner's panic grass is found in Canada from Ontario west to British Columbia and throughout the U.S., except South Carolina and Nevada. *D. oligosanthes* v. *oligosanthes* is found occasionally with a somewhat lesser area of occurrence. The varieties have small differences [19]. Specifically, v. *oligosanthes* has spikelets 3.2–4 mm long, grains 2.8–3.0 mm long, culms appressed-pubescent, first glume sparsely hirsute, and grains 1.5–1.6 mm broad. Variety *scribnerianum* has spikelets and grain length identical to v. *oligosanthes*, culms glabrous or spreading pubescent, first glume glabrous, grains 1.8–1.9 mm broad. *Dichanthelium oligosanthes* v. *scribnerianum* is found on mesic to dry soils in prairies or savannas; it thrives somewhat better with more friable soils. Although its Coefficient of Conservatism is only 3, it is not aggressive. It rarely competes with tallgrass, as it is more often found coexisting with forbs. The inflorescence appears in late May. The grains may be stripped from the seed-head when they appear tan in color in mid- to late June. The grains need no specific treatment for germination [11].

2.44 *Dodecatheon meadia* ssp. *meadia* (Primulaceae Family), Shooting Star

Dodecatheon meadia ssp. *meadia*, shooting star (height, 0.15–0.5 m), is an attractive early to mid-May flower found in savannas, woodland edges, and open prairies. A number of forms vary in flower color from white to pink to lilac (forma *album* is shown in the illustration on page 107). Preferred ecosystems are wet-mesic to dry-mesic soils; however, in drier situations shooting star is usually found along woodland edges. The typical subspecies *meadia* has its northern limit from New York west to Minnesota and Kansas. South of the northern limit it is found in every state, except New Jersey and Delaware. One other subspecies is largely restricted to the mid-south and Virginia. A Coefficient of Conservatism of 6 is indicative of its ease of restoration, provided that the proper ecosystem is selected. Oval seed capsules senesce into a brown color in early to mid-July, at which time the seed-head can be clipped. As the seed capsule has a hole in the top, care should be taken to hold the seed-head erect until bagged to prevent seed loss. Often high winds result in seed loss, so it is advisable to collect the seed-heads on time. After drying, the seeds (brown spheres about 1 mm in diameter) are easily thrashed out of the seed capsule. If complete recovery is desired, the capsules can be crushed. Partial seed purification can be achieved by sieving through window screen. The seeds of shooting star are best scattered in preferred areas in the late fall. Best success is obtained by sowing the seed along the forest prairie interface where partial shade occurs. If not fall planted, a 3-week cool-moist stratification followed sowing on top of the soil in the early spring (March) is recommended [11]. Longer stratification times will often result in seed germination in the refrigerator. These finding have been confirmed by germination studies [12]. That is, fresh seeds germinate 98% at 40°F in 8–16 weeks, and 97% in April if placed outdoors in September. Seeds dry stored 6 months at 70°F germinated 100% in 5–7 weeks when incubated under cool-moist (40°F) conditions [12]. The plant tolerates cool to cold weather incredibly well. When spring weather suddenly plunges into subfreezing temperatures, this plant's leaves actually will become stiff, but it will recover with subsequent warming with only slight initial leaf wilting.

**Dodecatheon amethystinum*, amethyst shooting star (height, 0.1–0.3 m, not shown), is similar to *D. meadia* described above, except that the plant is smaller (height, 0.1–0.3 m), and usually blooms about 1 week earlier, about the first of May. As the species name indicates that the flower is amethyst colored (deep red-purple), the color is certainly not a reliable indicator of the species. *D. amethystinum* has a white form (forma *margaritaceum*). Although the flowers of *D. meadia* are commonly found as lilac to pale pink in color, the *D. meadia* shown in the illustration is the white forma *album*. Except for one rare form, *D. meadia* leaves are reddish at their base, whereas the leaves of *D. amethystinum* are rarely so colored. Also, the flower petals (calyx lobes) of *D. meadia* number 3–7 (mostly 4–5), while *D. amethystinum* has 2–5 (mostly 3). In addition, *D. meadia* has firm dark reddish-brown seed capsules that are ovoid to barrel shaped, whereas the seed capsules of *D. amethystinum* have thin pale brown-yellowish cylindrical shapes. With a Coefficient of Conservatism of 9, amethyst shooting star is relatively difficult to restore. Although amethyst shooting star is found in the East (Pennsylvania and West Virginia), in the prairie states it is restricted to the "driftless area" of northern Illinois, southwestern Wisconsin, Iowa, and Minnesota. Restoration procedures follow the examples given above for *D. meadia*. The seeds are collected in late June to mid-July. As for *D. meadia*, the seeds are best scattered in the late fall, preferably along a woodland edge.

2.45 **Echinacea pallida* (Asteraceae Family), Pale Purple Coneflower

Echinacea pallida, pale purple coneflower (height, 0.6–1 m), is one of the more attractive of the prairie forbs, and it makes an excellent choice for landscaping of hillsides in full sun (illustration on page 107). Although it will grow on poorly drained soils, it will decrease over time. Pale purple coneflower prefers well-drained soil, such as mesic to dry sandy/gravelly soil or hillsides. However, it is somewhat sensitive to drought. It has been documented as being located in a northern limit from Maine west to Ontario, Wisconsin, Iowa, and Nebraska. Although it has been identified in most states south of its northern extent, it apparently is absent in several states of New England, the Middle Atlantic, Ohio, Kentucky, Mississippi, and Florida. Its so-called absence in Pennsylvania is belied by a prolific stand I have introduced into the south-central portion of the state. Although it has a high Coefficient of Conservatism of 7, it is relatively easy to restore provided the proper ecosystem is available. Flowers bloom in late June, and the dark-brown seed-heads can be clipped as early as late August. However, with the early harvested heads the seeds are extremely tightly attached, especially to the smaller seed-heads. The seeds lose their hold naturally as the season progresses, possibly by repeated exposure to rain. In fact, in particularly rainy seasons the seeds are loosened as early as late August. In normal to dry years seed-heads can be collected in late October to January. Collecting at later times results in some losses to birds and natural seed drop. Seeds will jar loose from larger seed-heads simply by thrashing, but even this method requires mechanical methods, especially with the smaller seed-heads. Depending on the weather, mid-October to November collection is an acceptable compromise. Certain seed suppliers use a device resembling a modified hand-cranked corn sheller. Although I have not tried using a leaf mulcher, the method would seem to have promise. Repeated pounding with the top portion of a sledge-hammer will break the seed-heads for subsequent seed removal by scraping over quarter-inch hardware cloth. After breaking the seed-head, twisting or scraping vigorously underfoot will often aid release of the seeds. Soaking the seed-heads in water and draining the excess water, followed by 2 days of sitting wet, will loosen the seeds, with some mechanical seed damage occurring during breaking by mechanical methods. If a limited number of seed-heads are available, one can shell the soaked seed-heads by hand with little damage to seeds. The seeds (tan-brown, wedge-shaped, 4–5 mm long) are best scattered on top of cultivated soil in late fall. Alternatively, a 90-day cool-moist stratification [11] followed by early spring scattering on top of either cultivated or uncultivated soil is recommended. To avoid expected weed problems, one can either sow the seed on uncultivated glyphosate-treated soil, or sow with annual rye on cultivated soil.

2.46 **Echinacea purpurea* (Asteraceae Family), Purple Coneflower

Echinacea purpurea, purple coneflower (height, 0.5–1.2 m), is another attractive forb that is grown in gardens everywhere (illustration on page 107). Unlike *E. pallida*, it prefers wet-mesic to mesic soils in partial shade. With more shade it will tolerate dry-mesic soil. Purple coneflower is found in every state south of its northern limit from Connecticut west to Ontario, Wisconsin, Iowa, Kansas, and Colorado, with the exception of New Mexico. A Coefficient of Conservatism of 6 is a good indication of its ease of restoration. Although mesic soils initially support the growth of purple coneflower in open prairies, increasing competition from other plants will cause its demise, but the plant will persist in its preferred niche along woodland edges. This so-called migration to edges is surprisingly rapid, probably because its lifetime is relatively short (3–5 years) and because it is a prolific seed producer. Purple coneflower has a better chance of surviving in full sun on rich wet-mesic soil. Inflorescence occurs in early July, and the seed-heads can be collected by clipping when senescent brown in late August to early October. In contrast to *E. pallida*, the seeds of *E. purpurea* are recovered far more easily. When the seed-heads are dry, the seeds (tan, wedge shaped, 4–5 mm long) will easily loosen with some thrashing or treading under foot. Scraping on quarter-inch hardware cloth sieve is also effective. Collection should be timely to avoid seed loss from bird predation or natural seed drop. When growing in the presence of *E. pallida* (hirsute, lanceolate-shaped leaves), the mostly glabrous-ovate leaves of *E. purpurea* give a quick check of the identity of the seed-head. Fall sowing on tilled soil surface is recommended by the author. Prairie Moon Nursery [11] claims that no seed treatment is necessary, but seed should be sowed on top of the soil in the spring. According to Deno [12], 6 months of dry storage at 70°F can be used to stimulate seed germination.

Unfortunately, many Nature Preserves have been "raided" for the root of purple coneflower, which is said to have curative powers as herbal medicine. In fact, roadside thievery of *Echinacea* has been noted on my property. Apparently, American Indians used the root even in prehistoric times. Research at the University of Munich reported the presence of an immuno-stimulating compound [20]. However, other studies have reported minimal to no medical effect on viral infections, such as the common cold. For example, in preventing or relieving the common cold, a 2005 article in the *New England Journal of Medicine* reported that those taking *Echinacea* fared no better than those taking a placebo.

2.47 **Elymus canadensis* (Poaceae Family), Canada Wild Rye

Elymus canadensis, Canada wild rye (height, 0.7–1.5 m), is a coarse grass of relatively low conservatism (Coefficient of Conservatism of 4). An early emergence of the grain-head is shown on page 107. Canada rye is found in most of Canada and the U.S., except the Southeast. The glaucous leaves often appear light blue-green. Its preferred ecosystem is mesic to dry soils in either prairies or savannas. It does well in sandy, gravelly, or rocky soils but will easily grow in many other soils. Canada wild rye produces a flowering seed head in mid-July and senesces to a tan color in mid-September to mid-October, when the seeds can be harvested by hand-stripping the seed-head or clipping the entire head. Seeds require no treatment for germination [11]. Fall planting is an option. This species is an early succession grass that tends to decrease when more conservative plants occupy its niche. If heavily seeded, it can be dominant in new restorations.

2.48 **Elymus villosus* (Poaceae Family), Silky Wild Rye

Elymus villosus, silky wild rye (height, 0.6–1 m), can be found in wet-mesic to dry-mesic situations mainly in partially shaded areas; however, it will survive in full sun. See page 107 for an illustration. It is found everywhere south of its northern limit from Massachusetts, Quebec, and Ontario west to North Dakota and Colorado; but it is absent in Louisiana and Florida. This grass affords diversity in restoration projects, but it is not particularly showy. The spiklets of the species are finely hirsute, which is a distinguishing feature. A Coefficient of Conservatism of 4 indicates a medium difficulty of restoration. Seed-heads should be gathered when first changing from greenish-tan to tan from late July to mid-August, after which the seeds shortly drop. Seeds can be readily recovered after a period of drying the seed-heads. Scraping over quarter-inch hardware cloth is an effective method. Seeds require no treatment for germination [11].

2.49 **Elymus virginicus* v. *virginicus* (Poaceae Family), Virginia Wild Rye

Elymus virginicus v. *virginicus*, Virginia wild rye (height, up to 1.5 m), grows in tufts in wet to mesic soils in partial shade to full sun throughout southern Canada and the U.S., except the West (illustration on page 107). One other variety is restricted to the northeastern and middle Atlantic states. Although Virginia wild rye has a wide distribution and is very easy to restore, it has a fairly high Coefficient of Conservatism of 4. Seed-heads can be collected in mid-August to early September.

Scraping on quarter-inch hardware cloth is an easy method of loosening the seed and partially separating the stems from the grain. The seeds require no treatment for germination [11]. Fall planting has been successful.

2.50 **Eryngium yuccifolium* v. *yuccifolium* (Apiaceae Family), Rattlesnake Master

Eryngium yuccifolium v. *yuccifolium*, rattlesnake master (height, 0.5–1.2 m), is unique with its yucca-like leaves, but without the rigidity associated with yucca (illustration on page 107). The typical variety *yuccifolium* is known to have a northern limit of Maryland and Ohio west to Michigan, Minnesota, and Nebraska. It is found south of the northern occurrence in all states, except Oklahoma and West Virginia. I have cultivated this plant very successfully in south-central Pennsylvania. Its relatively high Coefficient of Conservatism of 7 shows that it is not common. Its preferred habitat is on wet-mesic to dry-mesic soil in prairies or savannas. Although rattlesnake master grows readily on dry-mesic soil, it thrives better in wet-mesic conditions. It is fairly tolerant of shade as it is sometimes observed in shaded woodlands. Rodents occasionally do appear to eat portions of the root over winter, but generally the damage seems to be repaired the next growing season; in fact, rattlesnake master is relatively long lived if the habitat is proper. Plants will die in patches adjacent to healthy plants for no apparent reason, other than possibly lack of moisture. Green-white spherical flowers appear in early July, and the flower senesces to brown from late September to late October when the seed-heads can be clipped and bagged for drying. Loss of seed from natural seed drop is probable if seed-heads are not collected in a timely fashion. Processing of seed-deficient heads can give the illusion of seed collection, but most of the material primarily may be composed of bracts. Seeds (arrow shaped and scaly, 6 mm long) can be recovered from dried seed-heads by thrashing or scraping over quarter-inch hardware cloth. Two months of cool-moist stratification is required for germination [11]. Sowing seed in the late fall is a preferred method used by this author. Experiments showed that seeds germinated 84% in 8–11 weeks at 40°F [12], which implied that fall or early spring planting should be best. The growth of leaf rosettes of mature plants occurs very early in the spring. Rattlesnake master is easy to restore.

2.51 **Eupatorium altissimum* (Asteraceae Family), Tall Boneset

With a low Coefficient of Conservatism of 2, *Eupatorium altissimum*, tall boneset (height, 0.6–2 m), is a rather weedy species that volunteers in many restoration projects (illustration on page 107). Its northern limit is from Massachusetts west to Ontario, Minnesota, and Nebraska. From the northern limit it occurs in all states south. The small white flowers are not particularly showy. Its preferred ecosystem is full sun or partial shade in mesic to dry-mesic soil. Because it is a native species found in prairies, it is recommended that the plant be permitted to remain. As it tends to be an early succession plant, tall boneset will decrease naturally as your prairie matures. It is not recommended as a plant to actively restore unless diversity is desired. Seed-heads are collected from late September to late October. After drying, the seeds plus their pappus can be thrashed or scraped loose on quarter-inch hardware cloth. Seeds need 2 months of cool-moist stratification to germinate [11]. Also, fall sowing was very satisfactory in my experience.

2.52 **Eupatorium purpureum* v. *purpureum* (Asteraceae Family), Sweet Joe Pye Weed

Eupatorium purpureum v. *purpureum*, sweet Joe Pye weed (height, 1–2+ m) is found in wet-mesic to dry-mesic soils, mainly in partial shade (see page 107). The typical variety *purpureum* has its northern limit from Maine west to Ontario, Minnesota, and Kansas. It exists in all states south of the northern limit, except Texas. Another variety, *holzingeri*, is found in only seven states centered in Iowa and Missouri. Sweet Joe Pye weed has a medium Coefficient of Conservatism of 5. Its large size and light purple inflorescence appearing in mid- to late July lends variety to prairies or partially shaded areas. Seed-heads can be clipped and bagged when the seed-head becomes fluffy, about early August to early September. Seeds can be easily removed when the seed-heads become dry. The dark gray seeds are ridged longitudinally and measure 3–5 mm long by 0.5–1 mm wide with a pappus about 5 mm long. This author found that fall planting was successful. Controlled experiments with either fresh seed or seed dry stored for 6 months at 70 or 40°F gave 2% germination during 3 months of cool-moist treatment (40°F), followed by 98% germination at 70°F in 2–4 days [12]. Sweet Joe Pye weed has been easily cultivated by the author in full sun in mesic soil in weed-free beds.

Especially regarding immature plants, *E. purpureum* v. *purpureum* is not easy to distinguish from **Eupatoriadelphus maculatus* v. *maculatus*, spotted Joe Pye weed (not shown). In general, *E. purpureum* is slightly taller and has a pyramidal inflorescence versus a vertically flattened inflorescence of *E. maculatum*. Also, *E. purpureum* v. *purpureum* usually has a glaucous green stem with dark purple nodes [14]; *E. maculatum* has a rough stem either deep purple or spotted with purple. With *E. purpureum* v. *purpureum*, the individual flower heads are cylindrical, containing 3–7 florets subtended by four or five overlapping oblong blunt-tipped bracts of pink color; *E. maculatum* has individual noncylindrical flower heads containing 9–22 florets. Vigorously crushed leaves of *E. purpureum* v. *purpureum* have a sweet smell, described by some as a vanilla odor. In contrast to sweet Joe Pye weed, spotted Joe Pye weed has a preference for wet to wet-mesic soils.

Yet another Joe Pye weed, **Eupatoriadelphus fistulosus* (see herbarium scan on Springer internet site), is somewhat larger and also has a pyramidal inflorescence, but it is characterized as having a hollow stem. On the other hand, *E. purpureum* v. *purpureum* has a solid stem with pith intact in the center of a cross section. *Eupatoriadelphus fistulosus* is usually found in wet situations and generally in the shade.

2.53 **Euphorbia corollata* (Euphorbiaceae Family), Flowering Spurge

Euphorbia corollata, flowering spurge (height, 0.3–1 m), is highly branched when in flower (see illustration on page 108). This plant has a low Coefficient of Conservatism of 3, and it can thrive in ecosystems ranging from full sun to partial shade and mesic to dry soil. Its range is all states south of its northern limit of Maine west to Ontario, Minnesota, and South Dakota. It is often found in profusion on abandoned railroad beds where competition with other plants is less. The first flowers appear around the first of August, and seeds mature from the end of August to the end of September. The green fruit is a three-lobed capsule (about 4 mm) that protrudes above the flower petals. The great difficulty with restoring this species is seed collection. The three-lobed fruit should be picked when it just turns from green color to yellowish-green or tan. This procedure is basically opportunistic, as the seeds soon shatter from the capsule if not picked in a timely fashion. Worse yet, individual capsules ripen at widely varying times. Expect to obtain only a portion of the seeds. Capsules should be placed in a paper bag with the top closed, as the seeds shoot for some distance when the capsule splits open. Seeds are tan spherical ovals about 2 mm long. A large number of seeds are not necessary as this plant colonizes a site efficiently simply because the seeds are so ballistic. Seeds need a 1-month cool-moist stratification for germination [11].

2.54 **Euthamia graminifolia* v. *graminifolia, Solidago graminifolia* (Asteraceae Family) Grass-Leaved Goldenrod

Other than the thin leaves, *Euthamia graminifolia* v. *graminifolia*, grass-leaved goldenrod (height, 0.3–1.2 m), has an appearance typical of goldenrods (illustration on page 108). Its growth preference is either full sun or partial shade in wet to dry-mesic soils; however, it does prefer wet-mesic soils. It can be somewhat aggressive in these moist soils, spreading by rhizomes. The principal variety, *graminifolia*, is distributed widely over most of Canada and the northern U.S. but has sporadic presence in the South. It is absent in the Southwest, the Southeast, and four states in the Mid-South. One other variety is extremely limited to only Virginia and the Carolinas. Grass-leaved goldenrod has a low Coefficient of Conservatism of 3, indicating its ease of restoration. It flowers in early September, and the seed-heads can be clipped and placed in paper bags when the tops become fluffy, from late October to early November. When sufficiently dried, the seeds can be thrashed or scraped from the seed-head, giving very small seeds topped with a fluffy pappus. For germination the seeds should be given a 2-month cool-moist stratification followed by sowing on top of the soil [11].

2.55 **Fragaria virginiana* (Rosaceae Family), Virginia Strawberry

Fragaria virginiana, wild strawberry (height, 5–15 cm), has the aspect of ground cover with white flowers appearing in early May (illustration on page 108). In late May to early June when strawberries develop, the intense strawberry odor can be incredibly intense while strolling through large clusters of these plants. There are four subspecies, which are distributed over all of Canada and the U.S. *Fragaria virginiana* ssp. *grayana*, mainly found in the Midwest and a few eastern states, is distinguished by an ovoid to globose shape of the fruit (see the fruit in the Springer web site). Virginia strawberry enjoys a wide range of habitats from wet-mesic to dry soils in the full sun or partial shade. Although it is a desirable plant, it does prefer more disturbed areas (Coefficient of Conservatism of 2). Seeds found on the outer portion of the fruit require no treatment for germination [11]. This plant spreads by runners, and thus transplantation of a few plants can be an effective method of restoration. Although transplanting whole plants usually is not encouraged, the disruption of a few nonconservative plants is not a great loss. Where there is low competition, transplanted specimens readily form patches. The similar-appearing **Waldsteinia fragarioides*, barren strawberry, has yellow flowers, and its fruit is decidedly tasteless and odorless. Barren strawberry is rather common in the East but it is also found locally in the Midwest.

2.56 **Gaura biennis* (Onagraceae Family), Gaura or Butterfly Weed

Gaura biennis, gaura or butterfly weed (height, 1–2 m or taller), is a biennial plant rated at only 2 as its Coefficient of Conservatism (see illustration on page 108). Gaura usually is found in wet-mesic to dry-mesic prairies, especially if the area is more disturbed. Often it is found in disturbed dry situations, such as on abandoned railroad beds. Gaura is found in its northern range from Quebec west to Ontario and Minnesota. From its northern extent, it can be found in a majority of states south and east. However, it is absent in Florida, Missouri, Arkansas, Louisiana, and the upper New England states. The flowers, which appear in mid- to late August, change in color from white to reddish-pink as the flower ages. However, the plant tends to have a rangy appearance, which makes it less desirable as a horticultural specimen. The fruit containing the seed (6–8 mm long) can be collected from mid-September to early October. For germination, seeds require a 2-month cool-moist stratification [11] or fall planting. Restoration in new sites is very easy. Experimentally it was shown that 25% of seeds germinated in April after placing outdoors in October [12]. Without prior cool-moist treatment, 10% of the seeds germinated at 70°F in the light, whereas none germinated in the dark. Dry storage for 6 months had no effect on these results. The seed has been determined to have a half-life of 12 months [12].

2.57 **Gentiana alba, Gentiana flavida* (Gentianaceae Family), Cream Gentian

Gentiana alba, cream gentian (height, 0.3–0.9 m), has a cream-colored or whitish flower (illustration on page 108). The immature flower is closed, but the mature corolla opens, exposing a dentate pattern on top that appears as if it were gnawed. (See herbarium on Springer internet site; one flower of the herbarium has been cut open to reveal the inner structure.) *G. alba* tolerates a variety of soils (wet-mesic to dry-mesic), but it thrives in bottoms of draws in partial shade. It also does well in full sun. Its northern range extends from Pennsylvania west to Ontario, Manitoba, Minnesota, and Nebraska. The southern limit extends west from North Carolina to Kentucky, Arkansas, and Oklahoma. This plant is very easy to restore, which seems to contradict its Coefficient of Conservatism of 9. Cream gentian flowers in early September, and seed capsules can be collected from early October to mid-November. Seed capsules (elongated ovals about 2–3 cm long) can be either hand stripped or clipped from the plant. The seed capsules contain many oblong-flat seeds with wings (2–3 mm long) that are whitish tan. When the seed capsules are dry, they can be broken open simply by scraping across quarter-inch hardware cloth. A final sieving with screen removes extraneous matter and insect larva. Often maggot-like larva feed on the seed; thus, sieving with window screen is strongly recommended. It is possible that this insect pest may limit natural seed distribution by destroying the seed within the capsule. After processing, an immediate late fall sowing of the seed in established prairie has been highly successful in this author's experience. For seed germination, Prairie Moon Nursery recommends a 2-month cool-moist stratification followed by sowing on top of the soil [11]. Restoration of *G. alba* is much easier than either *Gentiana andrewsii* or *Gentiana puberulenta/puberula*. When grown in the presence of *G. puberulenta/puberula*, *G. alba* hybridizes with the former, giving intermediate forms, for example, cream-colored *G. puberulenta*-like flowers, blue flowers resembling *G. alba*, and intermediate whitish-blue and whitish-violet flowers. The *G. alba* × *G. puberulenta* hybrid has been named

G. × *curtisii* [15]. In my experience the hybrid appears as two basic forms, one with a white flower with purplish tint resembling the shape of *G. puberulenta* (see illustration on page 108) and the other a pale blue flower resembling the shape of *G. alba.*

According to Alan Wade of Prairie Moon Nursery (personal communication), *G. alba* and *Gentiana andrewsii* readily hybridize also. This hybrid has been named *G.* × *pallidocyanea* [15]. The hybridization of many *Gentiana* has been reported, including American with European species. Dr. Jan Gehring of Bradley University currently is investigating the hybrids of *G. alba* and *G. puberulenta* by genetic methods. Both *G. alba* and *G. puberulenta* compete well in the presence of the aggressive alien grass, *Bromus enermis* Hungarian brome, which is remarkable. Nonflowering *G. alba* can be distinguished from *G. puberulenta* by slightly larger green leaves and a yellowish-green stem, whereas the latter species has somewhat smaller purplish-green leaves and a purple stem.

**Gentiana andrewsii*, bottle or closed gentian (height, 0.3–0.8 m, not shown), is normally found as a blue flower that is completely closed at the top. Forcing open the closed petal reveals a fringed/dentate border (teeth are much smaller than in *G. alba*). A relatively high Coefficient of Conservatism of 7 shows that it is uncommon in disturbed areas. Bottle gentians grow in wet-mesic to mesic prairies or savannas. The flowers bloom in early September, and seed capsules can be collected in October. For germination, a 2-month cold-moist stratification is required, and seeds require sowing on top of the soil [11]. The white inflorescence of *G. andrewsii*, forma *albiflora*, can be confused with immature closed flowers of *G. alba.* As described above, *G. alba* flowers open when mature. Also, there are a number of additional anatomical differences between *G. alba* and *G. andrewsii.* For example, the lowest leaves of *G. andrewsii* are united bracts, which become gradually larger further up the stem as lanceolate to oblong-ovate, tapering at the end to a gradual point. *G. alba* leaves are uniformly lance-ovate, tapering at the end to a gradual point. Additionally, the leaves of *G. andrewsii* are thin/soft/pliable and somewhat leathery. The calyx-lobes (collection of sepals) of *G. andrewsii* are marginally fringed with hairs, whereas the sepals of *G. alba* are hairless, ovate/cordate, and spreading or turned downward.

2.58 **Gentiana puberulenta, Gentiana puberula* (Gentianaceae Family), Prairie or Downy Gentian

Gentiana puberulenta, prairie or downy gentian (height, 0.2–0.5 m), is truly a special sight with its large sky-blue blossoms nestled under the tall grasses of late autumn (illustration on page 108). Its habitat is mesic to dry prairies. Except for its recorded presence in New York, Maryland, and West Virginia, it is an exclusive inhabitant of the Midwest with a northern limit of Ontario west to Saskatchewan and North Dakota. Its limit in the south is Tennessee west to Arkansas, Louisiana, and Oklahoma. Downy gentian blooms in mid-September, about 1 week later than *G. alba*, and seed capsules can be harvested in mid-October to early November.

In the absence of the flower, *G. puberulenta* can be distinguished from *G. alba* mainly by the purplish color of the stem (see Sect. 2.57). Even though it has the same Coefficient of Conservatism of 9 as *G. alba*, downy gentian has earned its conservative rating. Seed harvest and treatment is the same as for *G. alba*. However, the seed of downy gentian is smaller and narrower (2 mm long), and the seed seems to be more susceptible to attack in the seed capsule by seed-feeding insect larvae. Seed loss is mainly caused by this insect pest. Success is obtained by sowing seed on established prairie areas in the late fall soon after seed harvest, particularly in areas of less dense growth. For seed germination Prairie Moon Nursery recommends a 2-month cool-moist stratification followed by sowing on top of the soil. Precise studies by Deno [12] demonstrated that seeds treated with cool-moist conditions (40°F) for 3 months gave 45% germination, followed by 24% germination at 70°F. He also showed that seeds sowed outdoors in January germinate in the spring. According to Deno (personal communication), the seedling roots of all the *Gentiana* are extremely sensitive to damage. With controlled germination on sterile soil, Deno has found that seedlings should be enclosed in plastic bags fastened loosely at the top. Perhaps this method mimics germination outdoors where tiny seedlings are nestled under prairie plants that would provide partial shade and moisture from transpiration. However, at the time of this writing Deno reports that only a few of his germinated seedlings had progressed beyond the dicotyledonous stage. When seeds are sown outdoors, expect plants to appear years later (at least 3 years). It is not clear whether the germinated plants are too small to be seen (probable), or seeds simply delay germination. Downy gentian is a must for prairie restorations. As already discussed, *G. puberulenta* is known to freely hybridize with both *G. andrewsii* and *G. alba* [15].

2.59 ***Geum triflorum* v. *triflorum* (Rosaceae Family), Prairie Smoke or Prairie Avens**

Geum triflorum v. *triflorum*, prairie smoke or prairie avens (height, 0.15–0.4 m), is one of the earliest blooming members of the prairie in mid-April, furnishing flowers with red petals and a yellow center (illustration on page 108). Its preferred ecohabitat is wet-mesic to dry sandy/gravelly soils in either full sun or partial shade. The typical variety of prairie smoke is found in more northern areas from New York and Ontario west and north to British Columbia, Northwest Territories, and Yukon. In the US Midwest it grows in Michigan, Wisconsin, northern Illinois, Minnesota, Iowa, and the Dakotas. In the West it is found in the three most northern states of the Rocky Mountains. Three other varieties are located only in the cooler regions of the West. This author has successfully obtained seed by culturing the plant in central Illinois in a raised bed of sand kept free of other plants. Even in northern prairies it has a high Coefficient of Conservatism of 9. The name prairie smoke is derived from the smoky appearance when in seed in late May. The seeds are attached to thread-like styles (2.5–5 cm long) that reflect light. Seeds require a

2-month cool-moist stratification for germination [11]. Because prairie smoke is a cool season plant, seeds should be sowed outdoors in the fall or early in the spring after cool-moist stratification. It is difficult to impossible to propagate in more southerly areas.

2.60 **Helianthus grosseserratus* (Asteraceae Family), Sawtooth Sunflower

Helianthus grosseserratus, sawtooth sunflower (height, 2–3+ m), prefers wet-mesic to mesic soils in full sun or partial shade (see illustration on page 109). Because the *Helianthus* genus has a fairly large number of species growing in a variety of situations, some defining characteristics for each are advisable. The mostly alternate leaves are lanceolate to oblong-ovate, tapering or rounded to the petiole. As the name implies, sawtooth sunflower has serrated leaves; however, individual plants often have very subdued serrations. Its stem is smooth and glaucous. As indicated by its Coefficient of Conservatism of 2, this plant is often seen in disturbed sites, such as roadsides. It is successful in a large region of the Canadian Ontario and Quebec, East, Midwest, and eastern Great Plains. It is absent in the southern states of South Carolina, Florida, and Alabama. Flowering occurs in early September, and seed-heads can be collected in mid-September to mid-October. After the seed-heads dry, seeds either can be thrashed loose, or rubbed over quarter-inch hardware-cloth. Seeds need no treatment for germination, or alternatively are stratified under cool-moist conditions for 2 months [11]. Because the plant can spread by rhizomes, it often can be very aggressive. It is not advised to restore this species unless diversity is important. If desired, it should be introduced only after a prairie has been firmly established, giving competition from other species. Thinning by uprooting plants may be necessary. Early emerging shoots are often eaten by unknown herbivores, but the plant readily recovers.

2.61 **Helianthus maximiliani* (Asteraceae Family), Maximillian's Sunflower

Helianthus maximiliani, Maximillian's sunflower (height, 0.5–3 m), is supposed to be native in states west of the Mississippi River, but this author has encountered it in an Illinois prairie in an area too remote to suspect introduction. See illustration on page 109. Current data show it is found in most southern provinces of Canada and all the U.S., except a few New England states, many states of the far West, and the Gulf Southeast. Alternate, rough lanceolate leaves that are folded upward along the midrib readily characterize Maximillian's sunflower. It grows in prairies from wet-mesic to dry-mesic soils. Although it seems to prefer drier sites, it is intolerant of drought. Because it spreads by rhizomes, some experts claim that it is too aggressive. If minimizing the number of seeds used in the restoration mix, this author believes that these claims of aggressiveness are much overstated. Prolific yellow flowers develop in late September, making it a good addition to the seasonal extension of a flowering prairie. Seed-heads can be clipped in mid- to late October. As goldfinches are fond of the seed, one should not wait too long. After drying the seed-heads, the seeds can be thrashed loose, giving typical sunflower-shaped brownish-black seeds 4–6 mm long. As for most *Helianthus* species, seeds can be either cool-moist stratified for 2 months or given no treatment for germination [11].

2.62 **Helianthus mollis* (Asteraceae Family), Hairy Sunflower

Helianthus mollis, hairy sunflower (height, 0.6–1.2 m), has distinctive ovate leaves that clasp the stem with scabrous (above) and pubescent (below) surfaces (illustration on page 109). Yellow flowers continue from early August to early September. It prefers mesic to dry-mesic soils, often sandy, in either open prairies or savannas. Even though this sunflower is tolerant of dry soil, conditions of drought cause a fair degree of stress. Its northern limit extends from Massachusetts west to Ontario, Wisconsin, Iowa, and Nebraska. It exists in all states south and east of the northern limit, except Florida. Hairy sunflower has a fairly high Coefficient of Conservatism of 7. Although this plant spreads by rhizome, it is not especially aggressive. It will slowly form dense patches of monoculture. Seed-heads can be clipped from early September to early October. Predation of seeds by the goldfinch can be expected. Seeds can be harvested by thrashing the seeds loose from dried seed-heads. Seeds are brown with a typical sunflower shape (3–4 mm long). For germination, the seeds require either no treatment or a 2-month cool-moist stratification [11].

2.63 *Helianthus occidentalis* ssp. *occidentalis* (Asteraceae Family), Western Sunflower

Helianthus occidentalis ssp. *occidentalis*, western sunflower (height, 0.4–1.5 m), is found in mesic to dry prairies and savannas (illustration on page 109). Western sunflower is an attractive sunflower characterized by its mostly basal ovate or oblong lanceolate leaves. This subspecies is found from New Jersey west to Minnesota and all states directly south of the northern limit, except Alabama and Mississippi. It is also found in Kansas and Texas. Another subspecies, *plantagu-ineus*, is only found in three states of the Mid-South. Although the Coefficient of Conservatism of western sunflower is a fairly high 7, it is easy to restore after treating seeds with a month of cool-moist stratification. Seed-heads can be clipped from late September to late October. After drying the seed-heads, the seeds are recovered by the method used for *H. grosseserratus*.

2.64 **Helianthus pauciflorus* ssp. *pauciflorus*, *Helianthus rigidus* (Asteraceae Family), Stiff Sunflower

Helianthus pauciflorus ssp. *pauciflorus*, stiff sunflower (height, 0.3–2 m), is a complex variable species with several races. It has been variously named *H. scaberrimus* [13] and *H. laetiflorus* v. *rigidus* [14], but another authority [15] assigned it distinctly as *H. rigidus* (currently renamed *H. pauciflorus*). The stem is hispid or scabrous (see illustration on page 109). The leaves are thick, coriaceous, serrate, and scabrous on both sides. The lower leaves are ovate or ovate-oblong and petioled, and the upper are lanceolate or oblong-lanceolate with a short petiole or sessile. The disk corollas have dark purple or brown lobes. Its habitat is mesic to dry-mesic prairies. Stiff sunflower has a northern range extending from Maine and Quebec west to Alberta and Montana. Its southern extent is Texas, Arkansas, and Kentucky. In the East and Southeast, stiff sunflower has a sporadic occurrence. With a Coefficient of Conservatism of 6, *H. pauciflorus* ssp. *pauciflorus* is tolerant of intermediate disturbance. Flowers appear in late August, and seed-heads can be collected in mid-October. Dried seed-heads can be treated by methods described for *H. grosseserratus* (above) to obtain sunflower-shaped brown seeds (5 mm long). Seeds require either no treatment or a 2-month cool-moist stratification for germination [11]. One other subspecies, *subrhomboideus*, has a slightly more western range compared to ssp. *pauciflorus*.

According to Barkley et al. [15], **H.* × *laetiflorus* (see herbarium on Springer internet site) is assigned as a persistent hybrid of *H. pauciflorus* and *H. tuberosus* (with consistently yellow disk florets). Its range is somewhat similar to the two parent species.

2.65 **Helianthus tuberosus* (Asteraceae Family), Jerusalem Artichoke

Helianthus tuberosus, Jerusalem artichoke (height, 1–3 m), undoubtedly derives its name from the edible tubers developed from enlarged rhizomes (illustration on page 109). This sunflower grows in wet-mesic to mesic soil in full sun or partial shade, but it prefers the more moist conditions of bottomlands. Jerusalem artichoke is widely distributed throughout the U.S., except the Southwest. In Canada it ranges from Quebec to Saskatchewan. Its Coefficient of Conservatism is a rather low 3. Because of its large size and multibranched top, it is a rather rough addition to a prairie, but the floral display can be strikingly attractive in the fall. The leaf characteristics are ovate or ovate-oblong, serrate, firm, three-nerved near the base, scabrous above, finely pubescent below. The upper leaves are alternate; the lower are opposite. Flowering occurs in mid- to late September, and seed-heads can be collected in mid- to late October. When the seed-heads are dry, sunflower-shaped seeds (5–7 mm long) can be obtained by the method described for *H. grosseserratus*.

Seeds require either no treatment or a 2-month cool-moist stratification for germination [11]. This author has observed that the seed-head can be rather sparse with viable seeds; it is likely that the enlarged rhizomes provide a major source of propagation. For example, cultivation often fragments the rhizomes and spreads them over a broader area, resulting in dense stands in agricultural fields. Transplanting tuberous rhizomes is an alternative method of propagation.

2.66 **Heliopsis helianthoides* (Asteraceae Family), False Sunflower or Ox-Eye

Heliopsis helianthoides, false sunflower or ox-eye (height, 0.8–1.5 m), is commonly found growing on wet-mesic–dry-mesic soil in prairies or savannas (illustration on page 109). There are four varieties: *gracilis*, *helianthoides*, *occidentalis*, and *scabra*. All overlap in their areas of occurrence, except *gracilis* occurring in the deep Southeast. The typical v. *helianthoides* is found north from Vermont west to Ontario, Wisconsin, and Missouri. It occupies every state south and east of the northern limit, except Florida. As indicated by its relatively low Coefficient of Conservatism of 4, it is often relatively common on recently restored sites, but decreases as the prairie matures. This sunflower is an attractive addition to new sites. Seed-heads can be collected from early August to early October. After drying the seed-head, the seeds are easily thrashed loose. Seeds drop easily from senescent heads; thus, regular collections are advisable. For germination, seeds can be either cool-moist stratified for 2 months or given no treatment [11]. In my hands, restoration by fall planting was entirely successful.

2.67 *Heuchera richardsonii* (Saxifragaceae Family), Prairie Alumroot

Heuchera richardsonii, prairie alumroot (height, 0.2–0.7 m), has a strong resemblance to the related garden-variety coralroot; however, the flower is yellowish green instead of coral in color (see illustration on page 110, including an inset showing a more mature flower). The species northern range is Ontario west to British Columbia and the Northwest Territory. In the U.S. it is found in the northern portion of the Midwest from Michigan and Indiana southwest to Oklahoma and northwest to Colorado, Wyoming, and Montana. The preferred ecosystems are prairies or savannas on wet-mesic to dry soils. Alumroot is often common on hillside prairies. Its Coefficient of Conservatism is a fairly high 7. It is a comparatively early flower in mid-May. Seed-heads can be clipped from early to mid-July. Seeds are easily shaken from the dried seed-heads. Seeds are very small (0.5–0.7 mm) black ovals. Because of their small size, seeds should be scattered on top of the soil, not covered with soil, and otherwise seeds require no treatment for germination [11]. This author has had success simply by sowing the seed immediately after harvest, and the following germination studies confirm this strategy. According to experiments [12], seeds germinated 92% at 70°F in 8–20 days, whereas cold-moist stratification followed by 70°F gave a total of only 42% germination. Light was not a factor. Seeds of this species have a half-life of only 12 months [12].

A similar species, **Heuchera americana*, is found more eastward. This species does not have the very hairy leaf stalks and leaf undersides possessed by *H. richardsonii*.

2.68 **Hypericum punctatum* (Clusiaceae Family), Dotted St. John's-Wort

Hypericum punctatum, dotted St. John's-wort (height, 0.5–1 m), is found in the north from Maine west to Quebec, Ontario, Minnesota, and Nebraska. It ranges south and east from this northern limit to all states. *H. punctatum* is identified by copious minute black dots found on the underside of the leaves and on the petals, requiring the use of a magnifying glass to view the dots. The white dots shown in the photo illustrated on page 110 are simply early morning dew. In overall appearance the plant somewhat resembles the aggressive alien, *Hypericum perforatum*. Both species possess black-dotted leaves, stems, and petals. *H. punctatum* has copious dots more or less evenly distributed over the underside of leaves, on the stems, and on the petals, whereas the leaves, stems, and petals of *H. perforatum* have far fewer black dots that are largely, but not entirely, found on the leaf underside margins. The restriction of the black dots entirely to the margins of the petals is the most characteristic feature of *H. perforatum*. The preferred habitat of *H. punctatum* is mesic to dry-mesic soils in full sun or partial shade. This species possesses a Coefficient of Conservatism of 3, indicating a fair degree of tolerance to disturbance. When senescent, the seed capsules turn black from mid-September to mid-October. After drying, the capsules should be crushed to release the very fine seed. Stems can be removed by sieving the seeds first through quarter-inch hardware cloth with a final cleanup with a window screen sieve. Germination can be stimulated by a 2-month cool-moist stratification followed by sowing on top of the soil [11]. This species also responds well to sowing the seeds in fall. This plant occasionally succumbs to disease, probably fungal infection.

2.69 **Hypoxis hirsuta* (Liliaceae Family), Yellow Star Grass

Hypoxis hirsuta, yellow star grass (height, 0.06–0.2 m), is a short plant resembling grass, but it is actually in the Liliaceae family. Attractive yellow star-like flowers appear in mid-May to early June (see illustration on page 110). The preferred restoration site is wet-mesic to dry-mesic soil in open woodlands, savannas, or rarely in prairies. It is usually seen on soil with some sandy/gravelly character. Its northern range is Maine west to Ontario, Saskatchewan, North Dakota, Colorado, and New Mexico. From this northern region it extends south to all states, except Florida. Yellow star grass is readily found on the lower slopes of the Eastern mountainsides. In the Midwest it is fairly rare and usually restricted to gravelly savanna hillsides. Although it has a Coefficient of Conservatism of only 6, it is not easy to restore. The fruit is an elliptical capsule 2–6 mm long that can be collected from early June to mid-July. Seeds are black (0.8–1.3 mm in diameter). For germination, 2 months of cool-moist stratification is required followed by sowing on top of the soil [11].

2.70 ***_Iliamna rivularis_ v. _rivularis_, _Iliamna remota_, and _I. corei_ (Malvaceae Family), Streambank Globemallow or Kankakee Mallow**

This plant, *Iliamna rivularis* v. *rivularis*, is included here because it is so interesting from the aspect of extreme separation of populations. *Iliamna rivularis*, streambank globemallow, was formerly thought to be restricted only to Western Mountain stream banks. Recently, the species has been divided into two varieties; see http://plants.usda.gov/java/nameSearch for information. The variety *rivularis* is found in the Northwest in Alberta west to British Columbia and south to Oregeon, Nevada, Utah, and Colorado. However, it is also known to be restricted to Altorf Island in the Kankakee River in Illinois, and as such, it was formerly categorized as Kankakee mallow, *Iliamna remota* (height, 0.6–1.7 m). See illustration on page 110. It was categorized as an Illinois endangered plant with a Coefficient of Conservatism of 10. Of coincidental interest, Altorf Island was the habitat of many rare plants, and is now part of an official Illinois Nature Preserve, known as the Kankakee River Nature Preserve. The existence of *Iliamna rivularis* v. *rivularis* (also known as *Iliamna remota*) eastward to Indiana and beyond probably has arisen from seed transport, such as along railroad right-of-ways. For example, a small colony is found along a railroad in Botetourt County, Virginia. Yet another population of *Iliamna rivularis* v. *rivularis* (also known as **Iliamna corei*) is found at the New River Narrows and Peters Mountain, Virginia. It is said that the so-called *Iliamna corei* was formerly restricted only to Peters Mountain. These widely separated populations are of considerable interest. Recently, nurseries sell Kankakee mallow/streambank globemallow, and thus it is fairly commonly cultivated. Kankakee mallow prefers wet-mesic to mesic soils in full sun, and it responds favorably to burn management. Some gravelly sandy character in the soil helps, which is similar to its original habitat, a dolomite bedrock river island or Western stream banks. Flowering occurs in mid-June, and seed capsules can be collected from late August to late September. Unfortunately, this plant attracts Japanese beetles, which can decimate the leaves and flowers. Seeds are borne in a very hairy mallow-like sphere (1.5 cm across). The hairy sphere breaks into papery quarter-moon segments about 1 cm long, which contain the seeds (2-mm spheres with an indentation).

Recommended seed treatment for germination is 2 months of cool-moist stratification [11]. Restoration is not easy; a small percentage of seed germination occurs after fall planting. A punctured seed coat is needed for efficient germination [12]. Because this mallow propagates from root rhizomes, it is possible to transplant from these colonies. Its nearest relative is *Iliamna rivularis* v. *diversa* (not shown), which also grows along streams in the northern Rocky Mountains in Montana, Wyoming, and Idaho. Coincidentally, I have observed *I. rivularis* in the western Rockies, and it appeared to me that the palmate leaves were more deeply cleft; thus, it is uncertain whether the variety was *rivularis* or *diversa*.

2.71 **Juncus tenuis* (Juncaceae Family), Path Rush

Juncus tenuis, path rush (height, 0.05–0.5 m), usually volunteers on paths where it receives little competition and tolerates treading under foot (illustration on page 110). Path rush has spread nearly worldwide. In the north it is found in much of Canada, including Alaska, and in all of the lower 48 states. A Coefficient of Conservatism of 0 seems a bit low, because path rush is only common where the competition is low or in wet-mesic soils. Flowering occurs in mid-June. The preferred ecosystem is wet-mesic to dry soils in full sun or partial shade. It is often found almost in monoculture in wet-mesic bottoms. Seeds gathered in mid-July are found contained within an oblong oval capsule. Oblong seeds are very small, 0.3–0.5 mm long. Seeds dry stored for 3 years germinated 90% at 70°F in the light in 6–10 days, but they did not germinate in the dark [12]. Thus, it is obvious that seeds should be sowed on top of the soil. Fresh seed germinated similar to the dry stored seed [12].

2.72 **Lespedeza capitata* (Fabaceae Family), Prairie Bush Clover

Lespedeza capitata, prairie bush clover (height, 0.6–1.6 m), is a legume that coexists well with tall prairie grasses. It consists of a leafy stem with few branches on top (illustration on page 110). Flower petals are an inconspicuous yellowish-white blooming around the middle of August. Its favored habitat is mesic to dry soils in full sun or partial shade. Prairie bush clover has a northern range from Maine west to Minnesota and South Dakota, and its southern extent is all states south and east of South Dakota. Prairie bush clover has a medium Coefficient of Conservatism of 4. Seed-heads can be clipped from the top of the plant in mid- to late October. Brown oval seedpods (4–5 mm long) can be thrashed loose when the seed-heads are permitted to dry. However, it is far easier to collect the seed pods in the field by bending the top of the plant over into a 5-gallon bucket and thrashing it against the bucket sides to loosen the pods, permitting the plant to remain largely intact. Inside the pod, one olive-brown oval seed (2.5 mm long) can be recovered. The procedure of rubbing with sandpaper effectively removes the pod as recommended for germination. After pod removal, the seed should be given a 10-day cool-moist stratification followed by inoculation with nitrifying bacteria specific for this species [11]. However, prairie bush clover is easy to restore by a simple fall planting of the seed with pod intact.

2.73 ******Lilium michiganense* (Liliaceae Family), Michigan Lily

Lilium michiganense, Michigan lily (height, 1–2 m), is one of the few prairie plants that produce spectacular orange-red to deep-orange flowers; the bloom occurs in early July (illustration on page 110). The usual ecosystem is wet to mesic soils in full sun or partial shade. It thrives marginally on mesic soil, preferring partial shade under these conditions. Michigan lily often forms colonies formed by the production of underground yellowish bulbs. Michigan lily is found along its northern range from New York west to Ontario, Minnesota, and South Dakota. It is absent in New England, and the eastern coast from New Jersey south. It is also absent in the mid southern states of Texas and Louisiana. Michigan lily's Coefficient of Conservatism is 6, but the restrictive moist habitat makes it a rather uncommon plant. According to Norman Deno (personal communication), one should be aware of two cultivation precautions. The first precaution is to remove tiger lilies from the vicinity as they carry a virus fatal to Michigan lily. The second is to either hand-pollinate or be certain that the natural pollinator, sphinx moths, are in the neighborhood. The pear-shaped to oval seed capsule (about 4 cm long) contain disk-like seeds that can be harvested from late August to early September. It is incorrectly assumed that the seeds require a warm-moist period followed by a period of cool-moist treatment [11]. By contrast, Deno [12] found that fresh seeds germinated 95% at 70°F. Dry storage 6 months at either 70 or 40°F resulted in 100% germination when cycled through 3 months moist 40°F followed by 70°F. However the dry stored seed was somewhat inhibited when exposed directly to 70°F, resulting in 83% germination (70°F dry stored) and 50% (40°F dry stored).

A related species, **Lilium superbum*, turk's-cap lily (not shown), is superficially similar in appearance, but it tends to have more yellowish flowers, ranging from yellow orange to orange. The leaves of Turk's-cap lily are smooth, whereas Michigan lily has minute spikes on the underside of the leaf along the margins and veins. Also, Michigan lily has somewhat smaller anthers (8–12 mm long), compared to Turk's-cap lily (17–25 mm long). Although *L. superbum* has the same requirements for habitat as *L. michiganense*, *L. superbum* seems to be more often encountered in partial shade in more southerly areas, including in the East. *L. superbum* has a Coefficient of Conservatism of 9 and is endangered in Illinois, being restricted to forests and stream banks in southern Illinois. Seeds of *L. superbum* germinate by treatments similar to those described for *L. michiganense* [12].

2.74 **Lithospermum canescens* (Boraginaceae Family), Hoary Puccoon

Lithospermum canescens, hoary puccoon (height, 0.1–0.45 m), is a rather small plant, which is found in the north from New York and Ontario west to Saskatchewan and North Dakota. It is found in all states south and east of its northern boundary, except Louisiana, Florida, and Delaware. Hoary puccoon makes up for its small size in mid-May with striking yellow flowers tinged with orange (see illustration on page 110). The name "hoary" is derived from its fine-hairy grayish-green leaves, and "puccoon" is an Indian word for plants that were used for dyes. Its habitat is mesic to dry soils in full sun or partial shade. Because hoary puccoon is often encountered either on hillsides with some sandy/gravelly character, old railroad beds, or black prairie silt-loam, loose friable soil is probably important. Some authorities claim that mycorrhizal associations are important with this species. Because this plant is usually found as isolated individuals, the Coefficient of Conservatism of 6 seems too low. However, it does tolerate a fair degree of disturbance. For example, in the photograph on page 110 it is associated with *Equisetum* sp., which is an indicator of disturbance. Because the plant fades into obscurity after flowering, the plants should be flagged for later seed collection. Smooth shiny-white nutlets (2–3 mm long) can be collected in August. This author does not know of a recommended seed treatment, but the structure of the seed is similar to those of the *Onosmodium* genus, which requires seed coat scarification, or more simply, fall planting. For the latter method, germination may require at least 2 years.

2.75 **Lobelia spicata* (Campanulaceae Family), Spiked Lobelia

Lobelia spicata, spiked lobelia (height, 0.2–1 m, usually less than 0.5 m), is an obscure plant that makes it presence known by its pale blue to white flowers in mid-June (illustration on page 110). Five varieties are known, of which four share most of the area. Spiked lobelia is distributed from Maine to North Dakota and south of the aforementioned states, except Florida. In Canada it is found from Quebec to Alberta. It prefers wet-mesic to dry-mesic soil in prairies or savannas. Although the Coefficient of Conservatism is a rather low 4, spiked lobelia does not respond well to disturbances such as mowing or grazing. Because the plants bearing seed are obscure, even when relatively abundant, it is best to flag the flowering plants for later seed collection. The seed capsules are hemispherical (3.5–5 mm in diameter), containing many brown seeds that are oval to elliptical (0.5 mm long). Depending on conditions, the seed-heads can senesce over a rather long period from late July to late October. Germination requires cool-moist stratification followed by sowing on top of the soil [11]. Fall sowing is also very satisfactory. The first year after germination results in rosettes of basal leaves with little to no likelihood of flowering.

2.76 **Lysimachia ciliata* (Primulaceae Family), Fringed Loosestrife

With a Coefficient of Conservatism of 4, *Lysimachia ciliata*, fringed loosestrife (height, 0.3–1.2 m), is usually found in wet-mesic meadows, but it will tolerate mesic soils (see illustration on page 111). Although fringed loosestrife is not very common, it is widely located in Canadian Quebec west to British Columbia and all the U.S., except the Southwest and Louisiana. Its yellow flowers with five petals bloom in mid-July. The reddish-brown three-angled seeds are contained in an oval capsule (diameter, 4–6.5 mm). Twenty-five to 40 seeds (about 2 mm in diameter) are found in each capsule. It is my experience that far fewer seeds are often found in each capsule. Recommended seed treatment for germination is a 2-month cool-moist stratification [11]. Deno [12] found 17% germination in the 12th week at 70°F without prior cool-moist treatment.

2.77 **Mirabilis nyctaginea* (Nyctaginaceae Family), Wild Four-O'clock

Mirabilis nyctaginea, wild four-o'clock (height, 0.3–0.8 m), furnishes a rare touch of reddish, pink, or purple to the prairie (illustration on page 111). According to Taft et al. [6], wild four-o'clock was not originally found in Illinois and eastward. Currently, it is adventive nearly everywhere; that is, the southern provinces of Canada and all the U.S., except Oregon, Arizona, and the southeastern coastal states. Its original range was Wisconsin west to Montana and south to Tennessee, Texas, New Mexico, and Mexico. Its soil preference is mesic to dry-mesic in full sun. Wild four-o'clock tolerates a fair amount of disturbance, and it can be rather weedy in certain soils. It is not particularly aggressive in mature prairies. It is easy to propagate and has a high probability of flowering the first year. The grayish-brown to nearly black fruit (4–6 mm long) is a hard oblong cylinder that is five-ribbed and angled. The fruit contains the oblong yellowish-brown seeds (3–3.5 mm long). Although wild four-o'clock produces attractive flowers in early June, green flowers appear much later. The later-blooming green flowers are more prolific, in bearing fruit about early September. Fruit collection plus debris can be accomplished by scraping the dried seed-heads over quarter-inch hardware cloth. Processing the fruit for seed germination is unnecessary, as no treatment is required [11].

2.78 **Monarda fistulosa* ssp. *fistulosa* (Lamiaceae Family), Wild Bergamot

Monarda fistulosa ssp. *fistulosa*, wild bergamot (height, 0.5–1.2 m), adds an attractive flush of pale purple to prairies or savannas during mid-July. An illustration is shown on page 111, and it is prominently displayed in a recent restoration in Fig. 5.2 with another early succession plant, *Ratibida pinnata*. The subspecies *fistulosa* is subdivided into five varieties, which are located in much of Canada and the U.S., except California and Florida. One other subspecies, *brevis*, is restricted only to the Virginias. Crushed leaves of wild bergamot give an odor reminiscent of oregano with a faint touch of mint. Also, the odor has been compared to Earl Grey tea, which contains oil of bergamot. As indicated by its Coefficient of Conservatism of 4, this plant thrives on wet-mesic to dry-mesic soil in new restorations, but it is an early succession plant that decreases as the prairie matures. Wild bergamot persists longer on wet-mesic and mesic soil. Seed-heads can be clipped from mid-September to mid-October. After drying in a paper bag, the seed-heads can be thrashed or scraped on hardware cloth to obtain oval brown seeds (about 1.5 mm long). Sieving through window screen provides greater purification. The seeds need no treatment for germination [11]. The plant readily spreads by rhizomes. According to Deno [12], germination was complex, but light was preferable. Seeds dry stored for 0, 6 months, 1 year, and 2 years germinated in the light at 70°F at 73, 80, 61, and 30%, respectively, whereas in the dark the germination was 0, 15, 26, and 25%, respectively. After 6 months storage of seeds at 70°F, germination was very good in either the light or dark provided that a 3-month cool-moist period was followed by a 70°F treatment. Seeds have a half-life of about 2 years.

2.79 **Oenothera biennis* (Onagraceae Family), Evening Primrose

Among the *Oenothera* genus, *Oenothera biennis*, evening primrose (height, 0.3–1.8 m), is the most common, being rather weedy and widely distributed throughout the U.S. and southern Canada, except the states of the Rocky Mountains. See illustration on page 111. As the species name *biennis* implies, it is a biennial that usually volunteers in restorations, but it decreases as the prairie matures. Evening primrose thrives in wet-mesic to dry soils in full sun or partial shade, particularly in disturbed situations (Coefficient of Conservation, 1). Although this plant is a native of prairies, it is not especially recommended to include in restorations, but rather volunteer plants can be tolerated. Flowers start appearing in mid- to late July. Because the flowers are open and most notable in early morning hours, one might rename the plant "morning primrose." Seeds are contained in cylindrical capsules tapered at the tip (1.5–3.3 cm long by 4–6 mm thick). The seeds are retained in the capsules even into the late fall; thus, seed-heads can be collected at any time after they become senescent brown. Seeds can be easily obtained by thrashing the seed-heads after drying, but more efficient removal of seeds may require crushing the seed-heads under foot. The angled reddish-brown seeds are 1.2–1.8 mm long by 0.8 mm broad. For germination seeds need no treatment, but preferably they should be sowed on top of the soil either in the fall or early spring to take advantage of a requirement for cool weather [11]. Experiments showed that the seeds require light to germinate; dark conditions gave no germination [12]. When provided with light at 70°F, seeds dry stored for 0 time, 6 months, 1 year, 2 years, and 3 years germinated within 3 weeks at 55, 60–75, 54, 13, and 14%, respectively. The best conditions for germination occurred after a 3-month cool-moist period followed by 70°F in the light, after which 85% germination was found for fresh seeds and 90–95% for seed dry stored for 6 months.

2.80 ***Oligoneuron rigidum v. rigidum, Solidago rigida* (Asteraceae Family), Rigid Goldenrod**

By its relatively large basal ovate/oval leaves and large flowers, *Oligoneuron rigidum*, rigid goldenrod (height, 0.3–1.5 m), is a rather unusual goldenrod in appearance (illustration on page 111). Its preferred ecosystem is full sun or partial shade in wet-mesic to dry soils. The typical variety *rigidum* extends its northern range from Massachusetts and New York west to Ontario, Minnesota, and Nebraska. From the north it is found in all states south and east, except four states in the Southeast. The variety *humile* is found in the Canadian and U.S. Midwest and the Great Plains to the Rocky Mountains. One other variety, *glabratum*, extends from Ohio south to Georgia and from Missouri south and west to Texas. As indicated by its Coefficient of Conservatism of 4, it is relatively easy to restore. Rigid goldenrod is not particularly aggressive, but it can be fairly prevalent in early succession prairies, decreasing over time. Flowering starts in early September, and seed-heads can be clipped for drying when the fluffy pappus becomes prominent from early to late October. After drying, the seeds plus pappus can be thrashed from the seed-head. The whitish seeds (3 mm long by 1 mm wide) are lined longitudinally with tan stripes and topped with a fluffy pappus about 4–5 mm long. For germination, the seeds should be given a 2-month cool-moist stratification followed by sowing in the spring [11], or seeds can be scattered in the late fall.

2.81 *Orbexilum onobrychis, Psoralea onobrychis (Fabaceae Family), French Grass, Sainfoin, or No Hay

Orbexilum onobrychis, French grass, which is not actually a grass (height, 1–1.5 m), is a member of the Fabaceae family. The leaves of French grass have a resemblance to those of soybean (illustration on page 111). Attractive blue to purple flowers can be seen in early July. French grass is found in mesic to dry-mesic soils, often on hillsides, in full sun or partial shade. Its northern range is from Virginia west to Ohio, Illinois, and Iowa. In the south its limit is South Carolina west to Tennessee and Arkansas. Its Coefficient of Conservatism is rated at 6. The one-seeded blackish pods (obliquely oval and corrugated, about 1 cm long) can be stripped from the seed-head in late September. For germination, seeds should be scarified and cool-moist stratified for 10 days [11]. This species has been observed to prematurely germinate in fall plantings; thus, spring planting is preferred. Once established, this plant readily spreads by rhizomes. Similar to many plants that spread by rhizomes, it rarely produces copious amounts of seeds.

2.82 **Oxalis violacea* (Oxalidaceae Family), Purple Oxalis or Violet Wood Sorrel

This delicate violet flower, *Oxalis violacea*, purple oxalis or violet wood sorrel (height, 0.1–0.25 m), is always a pleasure to find in mid-May (illustration on page 111). Purple oxalis is widely distributed from Massachusetts west to North Dakota and Wyoming, and is found south of the northern states to the U.S. border. It is also located in Oregon and Arizona. Under favorable conditions flowering also may reoccur in late August. Purple oxalis emerges from a bulb similar in appearance to a crocus bulb. It can be located in mesic to dry-mesic soils in prairies, savannas, or open woodlands. In shaded areas it can tolerate dry conditions. The Coefficient of Conservatism is 5. Because the plant is small, flagging the plant for collection of seed capsules in June is advised. Seeds are borne in oval capsules 4–6 mm long. The flattened seeds have a wrinkled appearance contained within a network. It is recommended to collect the capsules without seed separation (difficult). For germination, seed capsules can be sowed in the fall or alternatively stratified under cool-moist conditions for 2 months [11].

2.83 **Packera paupercula, Senecio pauperculus* (Asteraceae Family), Ragwort or Balsam Groundsel

Packera paupercula, ragwort or balsam groundsel (height, 0.1–0.5 m), blooms in early May (see illustration on page 111). Because of its relatively small size, it is rather inconspicuous. With a fairly low Coefficient of Conservatism of 3, it tolerates a wide range of conditions from wet-mesic to dry-mesic soils and full sun to partial shade. In the north it is found throughout Canada and Alaska. In the East, it occurs in every state east of the Mississippi River, except South Carolina. West of the Mississippi River it ranges north of Missouri and South Dakota, and it is found in the Rocky Mountain states. Ragwort is reportedly poisonous [11], possibly because of the alkaloid, senecionine. Consumption of senecionine by cattle from the related alien plant *Senecio vulgaris* was shown to be metabolized by the liver into toxic 4-hydroxy-2(*E*)-hexenal [21]. Seeds attached to a fluffy pappus can be collected from late June to early July. A 2-month cool-moist stratification is recommended treatment for germination [11].

2.84 *Panicum oligosanthes* see *Dichanthelium oligosanthes*

2.85 **Panicum virgatum* v. *virgatum* (Poaceae Family), Prairie Switchgrass

Because of its tolerance for wet-mesic to dry soils in full sun or partial shade, *Panicum virgatum* v. *virgatum*, prairie switchgrass (height, 0.8–1.8 m), is an easy grass to restore (illustration on page 112). The variety *virgatum* occurs widely from Quebec to Saskatchewan in Canada and all the U.S., except Maine and the West Coast. The other variety, *spissum*, is restricted to the East from Maryland north. Switchgrass in the photograph shown on page 112 is dripping with early morning dew. This grass has a relatively low Coefficient of Conservatism of 4. It is advisable to include only a very small amount of seed in a restoration project, as it can be aggressive in heavily seeded restorations. When it is restricted to one location in a mature prairie, it tends to maintain its position. Interestingly, switchgrass is common on the sandy soil of the Atlantic shore. The inflorescence can be observed in mid- to late September. When the seed-head begins to acquire a tan color, seeds can be stripped by hand from late August to late October; significant seed loss occurs at later times. The seed plus chaff is about 5 mm long, containing the grain (about 3 mm long). There is no reason to remove the chaff before sowing. Seeds require no treatment for germination. Germination at 70°F is 80% in 3–9 days, and 3 months cool-moist treatment followed by 70°F gave the same results in 2–6 days [12].

2.86 **Parthenium integrifolium* v. *integrifolium* (Asteraceae Family), American Feverfew or Wild Quinine

Parthenium integrifolium v. *integrifolium*, American feverfew or wild quinine (height, 0.4–1.2 m), is considered one indicator of undisturbed prairie, and consequently it has a high Coefficient of Conservatism of 8. See illustration on page 112.

Its habitat is prairies with mesic to dry-mesic soils. The northern range of wild quinine extends from Massachusetts and New York west to Minnesota and Kansas. It is documented south and east of its northern range, except New Jersey, Delaware, and Florida. The range of one other variety is limited to the southeast coastal states from Virginia south to South Carolina. The white flowers appear in late June. When the seed-head senesces to a brown color from late August to mid-October, it can be clipped for drying. The bracts plus seed (4–5 mm long) can be broken loose by scraping the seed-head back and forth on quarter-inch hardware cloth. The dark gray egg-shaped/arrow-shaped seed (3–4 mm long) is contained within the tan-colored bracts. It is not worthwhile to remove the bracts. For germination seeds require a 2-month cool-moist stratification [11]. Fall planting has proven to be highly successful. Rodents sometimes feed on the roots, which may account for a rather variable lifetime of the plant.

2.87 ****Pedicularis canadensis*** **ssp.** ***canadensis*** (Scrophulariaceae Family), Wood Betony or Lousewort

Pedicularis canadensis ssp. *canadensis*, wood betony (height, 0.15–0.4 m), blooms in early to mid-May (illustration on page 112). The subspecies *canadensis* is distributed from Maine to North Dakota and south, including much of the southern Canadian provinces. One other subspecies has been identified as being restricted to Colorado and New Mexico. Despite its Coefficient of Conservatism of 7, it is very difficult to restore. If the conditions are not correct, one can expect its disappearance. It is found on wet-mesic to dry soils mainly in savannas; however, it occasionally is found in prairies. It does best on well-drained soil with sandy/gravelly character or on hillsides. Wood betony is parasitic on other plants, which lends another degree of difficulty. In my opinion, a successful placement is sometimes unpredictable. During June, seed is borne in a lance-like oblong capsule 14–16 mm long and 4 mm wide. Seeds require a month of cool-moist stratification, and should be sowed in an established prairie so that host plants are available [11].

2.88 ****Penstemon digitalis*** (Scrophulariaceae Family), Foxglove Beard-Tongue

Penstemon digitalis, foxglove beard-tongue (height, 0.5–1 m), is usually highly successful in restoration projects (see illustration on page 112). This plant is an escape from its original habitat in the Mississippi Basin. Currently, it is found in all states south and east of a northern limit from Maine and Quebec west to Ontario, Minnesota, and South Dakota, but it is absent in Florida. A Coefficient of Conservation of 4 is indicative of a tolerance for disturbance. It lends a flush of white to prairies in mid-June at a time before the blossoming of most prairie species. Late into the fall and winter the reddish-purple basal leaves persist. Foxglove beard-tongue is tolerant of wet-mesic to dry-mesic soils in full sun or partial shade. The seed-heads are ready for clipping from early September to early October, and they should be permitted to dry in paper bags. Because the seed-head is composed of numerous open-ended capsules (8–14 mm long), relatively clean seeds can be obtained simply by thrashing. However, if a more thorough yield is desired, crushing the capsules underfoot will give seeds plus capsule pieces. An interesting odor reminiscent of unwashed feet may be entertaining during seed processing. Final seed purification by sieving through window screen affords reasonably clean seeds. The seeds are brown (0.7–1 mm). For germination the seeds can be sowed on top of the soil in the early spring after 1-month cool-moist stratification [11]. This author found that sowing on top of the soil in the fall was also highly successful. Deno [12] found that germination has an absolute requirement for light. Good germination occurred after dry storage for 6 months at 70 or 40°F, giving 90% germination at 70°F in the light (induction time, 4 days), but fresh seed gave only 26% germination at 70°F in the light. Dry-stored seeds responded well to a 3-month cool-moist treatment (40°F) followed by 70°F in the light (95–100% germination in 2–3 days).

A species somewhat similar in appearance, **Penstemon pallidus*, pale beard-tongue (not shown), mainly is found in sandy/gravelly soil ecosystems in the East and Midwest, except the deep Southeast.

2.89 *Petalostemum* see *Dalea*

2.90 **Phlox pilosa* ssp. *pilosa* (Polemoniaceae Family), Downy Phlox

Although *Phlox pilosa* ssp. *pilosa*, downy phlox (height, 0.3–0.6 m), will grow on soils ranging from wet-mesic to dry, this species prefers friable soils, such as those with either sand, gravel, or silty-loam character. See illustration on page 112. Downy phlox will grow in full sun or partial shade. There are a total of nine subspecies, of which eight subspecies have a fairly limited range. Subspecies *pilosa* has a wider distribution from its northern limit of Connecticut west to Ontario, Wisconsin, Iowa, and Kansas. The southern range is all states south of this northern limit. The plant blooms in mid-May, usually with pink flowers, but they can also be purplish to white. This is an attractive plant for restorations, but its Coefficient of Conservatism of 7 indicates its difficulty of cultivation. In the non-optimal soils, such as clay, it is not persistent in my hands. Because of seed shatter, seed collection is a challenge (see *Phlox bifida* in Chap. 3). Seeds become ripe in the latter part of July. According to Prairie Moon Nursery, seeds require a 2-month cool-moist stratification to germinate [11]. Deno [12] found that optimal germination of fresh seed is more complex, requiring several alternating 3-month cool-moist and warm-moist conditions. Germination was nearly complete with alternating 3-month periods of 70°F (7%), 40°F (7%), 70°F (7%), 40°F (27%), and 70°F (33% in 2–3 days). Using a 40°F period followed by 70°F, germination was only 17% in 6–9 days.

2.91 **Physalis heterophylla* v. *heterophylla* (Solanaceae Family), Clammy Ground Cherry

Physalis heterophylla v. *heterophylla*, clammy ground cherry (height, 0.45–0.8 m), is in the Solanaceae family, essentially one of many species of wild tomato (illustration on page 112). It tends to be a weedy species and thus will thrive in newly established prairies or savannas with mesic to dry-mesic soils. It is an early succession plant with a Coefficient of Conservatism of 2. Clammy ground cherry is found throughout the U.S., including Canadian Quebec and Ontario, but it is absent in Nevada and California. The yellowish oval seeds (2–2.5 mm long) are contained within a yellow fruit (10–12 mm long). I am aware of no method to promote germination. Deno [12] found that other members of the *Physalis* genus required seed removal from the fruit followed by washing with water (three rinses of water per day for 1–4 weeks). Using washed seed, light was required for germination. One other variety is restricted to Texas.

2.92 *Physostegia parviflora* (Lamiaceae Family), Western Obedient Plant

As noted by the square stems, *Physostegia* are members of the mint family. *Physostegia parviflora*, western obedient plant (height, 0.3–0.9 m), is very similar in overall aspect to another member of the mint family, *Teucrium canadense*, except the leaves of western obedient plant are comparatively narrower (illustration on page 112). It is a wet to mesic soil prairie plant that is found in the northern half of the U.S. in the Midwest and West, as well as mid- and western Canada. Western obedient plant is a rather rough species that spreads in colonies by rhizomes fairly aggressively. The flowers bloom in early to mid-July, and each floral bract contains four oblong brown nutlets 2–3.3 mm long. Seeds in the lower bracts ripen first, about the first of August, even while the top of the floral stem continues to bloom. It is preferable to tolerate loss to seed drop from lower capsules by simply cutting the seed-head near the end of flowering in mid- to late September. The seeds can be jarred loose from the seed-head by scraping over quarter-inch hardware cloth;

expect much extraneous material mixed with the seeds. For germination the seeds require a 2-month cool-moist stratification [11]. Transplants from spreading rhizomes are also possible.

2.93 **Physostegia virginiana* ssp. *virginiana* (Lamiaceae Family), Obedient Plant

Compared to *P. parviflora* described in Sect. 2.92, *Physostegia virginiana* ssp. *virginiana* (obedient plant) is somewhat taller (0.4–1.5 m tall) and has much larger flowers (terminal spike, 6–20 cm long; corolla, 2–3 cm long) (see illustration on page 112), in contrast with the floral size of *P. parviflora* (terminal spike, 3–15 cm long; corolla, 0.8–1.5 cm long). The appearance of *P. virginiana* is notably more attractive than *P. parviflora*. *P. virginiana* has a Coefficient of Conservatism of 6. The preferred habitat of *P. virginiana* is wet to mesic soils in full sun. Occasionally, it can be found on dry-mesic soils. Obedient plant thrives best in wet-mesic bottoms. It is distributed from Maine west to North Dakota and south of this U.S. northern limit, except for Texas, Arkansas, Louisiana, and Florida. It is also located in Montana, Utah, and Canadian Quebec west to Manitoba. Obedient plant spreads by rhizomes but is less aggressive compared to the previously described *P. parviflora*. Showy pink to purple flowers appear in mid-August, and the seeds can be collected in early October to mid-November when the seed-head senesces. See Sect. 2.92 for seed processing. Prairie Moon Nursery [11] recommends a 2-month cool-moist stratification for seed germination. However, Deno [12] found that at 70°F seed germination was 70% complete in 4–6 days, whereas a 3-month cool-moist (40°F) period followed by 70°F gave only 43% germination in 3–5 days.

2.94 **Polemonium reptans* v. *reptans* (Polemoniaceae Family), Jacob's Ladder

Polemonium reptans v. *reptans*, Jacob's ladder (height, 0.15–0.5 m), is found in a broad range of conditions from wet-mesic to dry-mesic soils, mainly in open woodlands, and in savannas (illustration on page 113). This species is occasionally found in full sunlight in prairies, and it seems to thrive in the open provided that competition from other plants is not too severe. Because of its attractive flower and early blooming in late April to early May, Jacob's ladder is included here. The variety *reptans* has a northern limit from Quebec west to Ontario, Minnesota, and South Dakota. In the south and east of the northern range it is located in all states, except Maine, Nebraska, Texas, Louisiana, Florida, and South Carolina. One other variety, *villosum*, is restricted only to Ohio and Kentucky. The Coefficient of Conservatism is rated at 5. About three seeds are contained in each capsule. The seeds ripen at various times in early June, and they shatter rather easily. Clipping the seed-heads before capsule shatter is recommended; that is, when the capsules just turn from green to yellow or tan. Seeds can be cleaned by passage through a window screen. According to Prairie Moon Nursery [11], seeds require 2 months of cool-moist stratification for germination. Fall planting has proven to be somewhat successful in my hands. Germination experiments [12] showed that multiple alternating 3-month cool-moist (40°F) and warm-moist (70°F) periods with fresh seeds are best; dry storage is harmful to the seeds. Accordingly, 40, 70, and 40°F gave 44% germination in 9–12 weeks, and additional 70 and 40°F treatments gave 11% more germination. Judging from numerous volunteer plants surrounding beds of *P. reptans* v. *reptans*, it is this author's opinion that seeds should be planted/scattered immediately or very shortly after collection, which would somewhat mimic the conditions described by Deno [12] above. It is my observation that the plants are not obvious the first or second growing season.

2.95 **Polygala sanguinea* (Polygalaceae Family), Field Milkwort

Polygala sanguinea, field milkwort (height, 0.1–0.4 m), is one of the few annual plants found on the prairie (illustration on page 113). It is usually found on prairies composed of rather sterile mesic to dry-mesic soils, which are often acidic. This plant probably cannot tolerate much competition. Its northern limit extends from Nova Scotia west to Ontario, Minnesota, South Dakota, and New Mexico. It is found in all states south and east, except Florida. Its Coefficient of Conservatism is rated at 5. Because the fruit drop promptly, it is important to collect them on a timely basis as the fruit ripen. The petals/fruit are collected starting at the bottom when the whitish flower bracts easily come loose in the hand. The time of fruit collection is during the first half of August when the flower is at the peak of bloom. The seed is oblong/oval with short hairs (1.5–1.7 mm long). If the fruit plus bracts are scattered shortly after collection in a relatively barren area, the restoration of this plant is generally successful.

2.96 **Polygonatum biflorum* v. *commutatum, Polygonatum commutatum/canaliculatum* (Liliaceae Family), Great Solomon's Seal

Polygonatum biflorum v. *commutatum*, great Solomon's seal (height, 0.6–2 m), is commonly found in open woodlands or savannas; that is, most often in more shaded areas. It can be difficult to establish in a full-sun prairie. Because of the unusual appearance of this graceful plant, it is included here (illustration on page 113). The species name *biflorum* hardly applies to this variety, as the flowers are generally found in clusters of three (occasionally two). The variety *commutatum* is one of five varieties; this particular variety is located from Maine west to Montana and south to the border. It also exists in Canadian provinces from Quebec west to Saskatchewan. It prefers mesic to dry mesic soils, but with increased sunlight more soil moisture is probably required. Its Coefficient of Conservatism is 4. Several seeds (3–4.5 mm in diameter) are borne in a cluster of dark blue berries (7–15 mm diameter) that can be collected in the latter part of August. There are two alternatives for seed germination. This author has successfully obtained germination by planting fresh seeds in full sunlight and waiting 2 years for germination. However, the plants soon disappeared, probably because the mesic soil in full sun was not conducive to its survival. Another recommended method requires 2- to 3-month periods of alternating conditions of cool-moist, warm-moist, and cool-moist [11]. Also, fresh seed should be used. Deno [12] found that a different species, *Polygonatum humile*, germinated after 10 weeks at 70°F in the dark, but developed an extensive root system before the stem and leaves appeared. Seedlings did not appear above ground until further 3-month cycles of 40, 70, 40, and 70°F had occurred (a full year of cycles). By my observation, such a pattern seems plausible for *Polygonatum biflorum* v. *commutatum*. Great Solomon's seal also spreads by root rhizomes.

2.97 *Polytaenia nuttallii* (Apiaceae Family), Prairie Parsley

Polytaenia nuttallii, prairie parsley (height, 0.4–1 m), superficially resembles the flower of the aggressive alien *Pastinaca sativa* (wild parsnip), but the difference in the leaf shape is the feature that most distinguishes the two (see illustration on page 113). *P. sativa* has pinnately compound cut-toothed leaves, whereas *P. nuttallii* leaves are twice pinnate (the segments cuneate and incised; see herbarium on Springer internet site). *Do not mistake this rare plant for parsnip***.** Flowers appear in late May, about three growing seasons after germination. This yellow-flowered member of the carrot family (Apiaceae; authorities disagree) has a relatively high Coefficient of Conservatism of 8. Prairie parsley is a typical Midwestern inhabitant with its northern limit from Michigan west to North Dakota. Its eastern limit is Alabama north to Tennessee, Kentucky, and Indiana. The western limit is New Mexico north to Kansas and Nebraska. Various authorities assign a soil ecotype of growth preference as wet-mesic to dry-mesic or simply dry. This author has successfully grown it in poorly drained mesic soil and dry-mesic shale soil. Prairie parsley will grow in either full sun or partial shade. Most authorities list the plant as a perennial, but one marked individual observed by this author succumbed after setting seeds; however, it did require 3 years before it matured sufficiently to flower. If several other members of the Apiaceae family are instructive, the plant may naturally succumb after setting seeds. The large distinctive seeds are actually classified as fruit. The seed/fruit are flattened ovals (5–10 mm long by 4–7 mm wide) with corky wings on the margins. The seeds can be collected from the umbel seed-head during the first half of August. The seed-head resembles that of *P. sativa*. Because the leaves of prairie parsley usually have senesced when in the seed stage, it is even more difficult to distinguish it from the parsnip seed-head. It is strongly recommended to flag the plant when it still has easily identifiable leaves to aid later collection. For germination seeds require 4 months of cool-moist stratification or fall planting [11]. Seeds may require more than a year to germinate.

2.98 **Potentilla arguta* ssp. *arguta* (Rosaceae Family), Prairie or Tall Cinquefoil

Although *Potentilla arguta* ssp. *arguta*, prairie cinquefoil (height, 0.3–1 m), is not a strikingly beautiful plant or flower, it ranks a high 10 in its Coefficient of Conservatism (illustration on page 113). In my hands it is so easy to restore that the assigned Coefficient of Conservatism seems much too high. Prairie cinquefoil responds very well to cultivation as a row crop or a weeded bed crop. It has a lifetime of at least a decade. Its preferred ecosystem is mesic to dry soils in full sun or partial shade. Its northern range is Maine, New Brunswick, and Quebec west to Alberta, and the Northwest Territory. It is found extensively in the U.S. with the exception of the entire southern region, that is, south of Virginia, Tennessee, Arkansas, Oklahoma, Colorado, and Idaho. One other subspecies *convallaria* is found in the Rocky Mountains and further west. Animals, probably deer, occasionally graze prairie cinquefoil, but the plant will recover. Flowering occurs in early July, and the seed-heads senesce to a brownish-tan color from early August to early September. The stem containing multiple seed capsules should be clipped carefully and kept upright until placed in bags, as seeds will spill when tipped over or jarred. After drying, tiny brown seeds (pear shaped with a slight beak at the top, 1 mm long) can be easily thrashed out of the seed heads. If a more thorough harvest is desired, the seed-heads can be scraped over hardware cloth. For further purification the seeds can be sieved through an ordinary window screen. The small seeds should be sowed on top of the soil either in the late fall, or in the early spring after treating with a 2-month cool-moist stratification [11]. This author had excellent success sowing seeds in the fall on top of relatively weed-free soil. In the East, the plant seems to thrive best on dry-mesic shale-laden soils.

2.99 **Potentilla simplex* (Rosaceae Family), Old-Field Cinquefoil

Potentilla simplex, old-field cinquefoil (initially erect up to 0.2–0.5 m, then spreading by runners to 1.2 m), is easy to establish as it roots at the tips of the runners (illustration on page 113). As indicated by its Coefficient of Conservatism of only 3, this plant is easy to establish. The habitat of *P. simplex* is mesic to dry prairie hillsides or savannas, often being found growing in sandy soil. Its northern limit is Newfoundland west to Ontario, Minnesota, and Nebraska. From its northern limitation it is found south and east in all states, except Florida. The yellow flowers bloom in late May. Although the principal mode of propagation is undoubtedly by rooting runners, seed capsules can be collected. The seed capsules tend to shatter readily. Yellowish-brown ribbed seeds are 0.9–1.2 mm long. Seeds should be spread on top of the soil in the fall.

A similar, but smaller plant, **Potentilla canadensis* (see herbarium on Springer website), is also a native of the prairie. One of the more easily identifying characteristics is the first flower of *P. canadensis* develops usually from the node above the first internode, whereas the first flower of *P. simplex* is usually found at the node above the second internode.

2.100 *Psoralea* see *Orbexilum* and *Psoralidium*

2.101 **Pycnanthemum tenuifolium* (Lamiaceae Family), Slender Mountain Mint

Pycnanthemum tenuifolium, slender mountain mint (height, 0.3–1 m), possesses narrower leaves (1.5–4 cm long by 0.1–0.45 cm wide) than both *P. verticillatum* v. *pilosum* and *P. virginianum*, and the crushed leaves give only a subtle mint odor compared with *P. verticillatum* v. *pilosum*. Its thin leaves give a notable diverse appearance to restoration projects (illustration on page 113). The preferred habitat is wet-mesic to dry-mesic soils in full sun or partial shade, but it does thrive somewhat better in wet-mesic and mesic soils. The plant has a tendency to spread by rhizomes. Its northern range extends from Maine and Quebec west to Ontario, Minnesota, and Nebraska. It exists in all states south and east of the northern limit. Its Coefficient of Conservatism is rated at 4. Slender mountain mint blooms in early July, and the seed-heads senesce into a blackish color from late August to late October. Seed harvest and seed germination follows the procedure described below for *P. verticillatum* v. *pilosum*. The oblong dark brown seeds or nutlets are somewhat smaller than those of the other *Pycnanthemum* species described (0.7–1 mm long).

2.102 **Pycnanthemum verticillatum* v. *pilosum*, *Pycnanthemum pilosum* (Lamiaceae Family), Hairy Mountain Mint

The leaves of *Pycnanthemumve verticillatum* v. *pilosum*, hairy mountain mint (height, 0.6–1.2 m), when crushed give a very intense and pleasant odor of mint. See illustration on page 113. Hairy mountain mint gives the strongest odor of all the other *Pycnanthemum* species described. Because it is known that mint odors are deposited into protrusions on the leaf surface, perhaps this is the function of the "hairy" leaf and stem. The plant propagates by rhizomes, but the multiplication of the plant by rhizomes is barely notable. Its preferred habitat is mesic to dry-mesic soils in full sun or partial shade. Its range extends into Quebec and Ontario in Canada. In the U.S. its northern limit is Massachusetts and New York west to Michigan, Illinois, Iowa, and Nebraska. It is absent in the Mid-South and much of the Southeast. Its Coefficient of Conservatism is 6. Hairy mountain mint blooms about mid-July, and the seed-head senesces to a gray color from mid-September to the end of October. When gray, the seed-head can be clipped and stored until dry. Thrashing or shaking will loosen many seeds, but more strenuous methods, such as scraping on hardware cloth, may be required to improve the yield, adding more trash material. If desired, more seed purification can be obtained by sieving through a window screen. The oblong seed or nutlets are brown (1–1.2 mm long). The seeds require no treatment for germination [11]. Because of their small size, the seeds should be sowed on top of the soil. Fall planting is an alternative method. *Pycnanthemum verticillatum* v. *pilosum* can be easily confused with *Pycnanthemum virginianum* (below). To distinguish the two species, several characteristics can be used. The sides and angles of stems of *P. verticillatum* v. *pilosum* are hoary-pilose (covered with fine white or gray thin hairs), whereas the sides of stems of *P. virginianum* are smooth, but the stem angles are softly hairy. *P. verticillatum* v. *pilosum* has shallowly (barely) dentate leaves that are hoary-pilose and lanceolate (2–7.5 cm long by 0.4–2 cm wide), whereas the leaves of *P. virginianum* are more smooth (sometimes sparsely pubescent below), linear-lanceolate, and generally narrower (2.5–5.5 cm long by 0.2–1.1 cm wide).

2.103 **Pycnanthemum virginianum* (Lamiaceae Family), Virginia or Common Mountain Mint

Pycnanthemum virginianum, Virginia mountain mint (height, 0.4–1 m) is similar in appearance, restoration methods, and growth habit to *P. verticillatum* v. *pilosum* described above; however, *P. virginianum* prefers a more moist habitat of wet to mesic soils. Distinguishing features to identify the two species are outlined in the Sect. 2.102 (illustration on page 114). Its northern range extends from New Brunswick west to Ontario, Minnesota, and North Dakota. From the northern limit it is found in all states south and east, except Texas, Louisiana, Florida, and South Carolina. Its Coefficient of Conservatism is rated at 6. Senescent gray seed-heads ripen from early September to late October. The brown seeds or nutlets are oblong and measure 0.8–1.2 mm long. See the Sect. 2.102 for seed recovery. The seeds require no treatment for germination, and they should be sowed on top of the soil [11]. In my hands, fall plantings gave results, but some plants appeared during the second springtime. In view of these latter observations, the "no-treatment" alternative seems to be suspect or the newly germinated plants develop slowly.

2.104 **Ratibida pinnata* (Asteraceae Family), Drooping, Yellow, Prairie, or Gray-Headed Coneflower

Ratibida pinnata, drooping coneflower (height, 0.5–1.5 m), may qualify as a species with the greatest number of common names, possibly because it is relatively common with notable yellow flower in prairies and savannas (illustration on page 114). Its northern range extends from Vermont and New York west to Ontario, Minnesota, and South Dakota. From this northern range it is found south and east in all states, except Texas, North Carolina, Maryland and Delaware. As its Coefficient of Conservatism of 4 indicates, it is usually an indicator of moderately disturbed prairies. Drooping coneflower is an easy plant to grow, and its attractive aspect makes it an excellent choice for suburban gardens. This author was surprised to find a single specimen in the Bear Meadows Natural Area, a remote area in the mountains near State College, Pennsylvania. The approximate lifetime of the plant is a decade, and this plant decreases as the prairie matures. The preferred ecosystem is mesic to dry-mesic soils in prairies and savannas. Flowering usually occurs in early to late July, and the seed-head matures (indicated by a brown stem) from mid-August to late September. The mature seed-heads can be clipped and dried in paper bags. After drying the seed-heads can be broken by vigorous treading and scraping underfoot. The more mature heads are the easiest to crush, and smaller and less mature seed-heads may require pounding with the top end of a sledgehammer. This crushing process gives an interesting odor somewhat resembling pine resin. Recovered from this method are blackish-gray seeds (about 3 mm long) that resemble small sunflower seeds. Also recovered are fines that are not worth separating from the seeds. However, many of the stems can be removed by sieving the seeds through quarter-inch hardware cloth. The recommended seed treatments for germination are either no treatment or a 2-month cool-moist stratification [11]. This author prefers the ease and success of fall planting.

2.105 **Rosa arkansana* v. *suffulta*, *Rosa suffulta/pratincola* (Rosaceae Family), Sunshine Rose, Arkansas Rose, or Prairie Wild Rose

Similar to most wild rose species, *Rosa arkansana* v. *suffulta*, sunshine rose or Arkansas rose (height, 0.2–0.5 m) has pinnately compound leaves with mainly 7 leaflets or occasionally 3–11 leaflets (illustration on page 114). *R. arkansana* possesses straight, very slender prickles that densely cover the entire stem, which distinguishes it from the overall appearance of *R. carolina* or *R. virginiana*. *R. carolina* has scattered, short, needle-like prickles toward the base of the plant, but the middle and upper stem has few or no prickles. *R. virginiana* is armed with thick-based straight or slightly curved prickles on both sides of the stem just below the attachment of the compound leaf (infrastipular). *R. arkansana* has been classified as two varieties: v. *arkansana* is found mainly in Canadian Manitoba west to British Columbia and in the U.S. Great Plains, also including Illinois and Wisconsin, whereas *v. *suffulta* (also known as *Rosa pratincola* or *Rosa suffulta*) is found slightly more eastward. Variety *suffulta* has a northern limit from Massachusetts and New York west to Ontario, Alberta, and Montana. In the south the range is north of the Ohio River and all states west of the Mississippi River to the Rocky Mountains, except for its absence in Louisiana. The difference in the two varieties concerns the smooth surface on the underside of the leaflets of *R. arkansana* v. *arkansana* versus minute downy hairs on the underside of the leaflets of *R. arkansana* v. *suffulta*. As a further complication, *R. arkansana* and *R. carolina* usually are found at sites together where they freely hybridize. *R. arkansana* grows in wet-mesic to dry soils in full sun to partial shade, where it readily spreads by rhizomes. *Rosa arkansana* v. *suffulta* has a Coefficient of Conservatism of 5. Seeds can be recovered from the red rose hips in the fall. The seeds should be scarified and given a 2-month cool-moist stratification; however, germination may occur in the second year or later [11].

2.106 **Rosa carolina* v. *carolina* (Rosaceae Family), Pasture Rose

Rosa carolina v. *carolina*, pasture rose (height, 0.2–0.9 m), is one of several native rose species (illustration on page 114). *Rosa carolina* v. *carolina* is widely distributed from a northern limit of Maine west to Quebec, Ontario, Minnesota, and Nebraska, and this typical variety is found in all states south and east of its northern reach. Another variety, *setigera*, is limited to New England. This latter variety should not be confused with the species *Rosa setigera*, which is large bushy plant with large recurved thorns throughout the branches. Also, *R. setigera*, described below, has usually three leaflets, whereas *R. carolina* has pinnately compound leaves composed of five to nine leaflets (usually seven). Occasional stout prickles curved downward are found mainly at the base of the stem internodes of *R. carolina* but seldom higher on the stem. When evident, small barely differentiated prickles, either straight or slightly bent downward, are found immediately below the attachment of the leaf with the stem. Pasture rose prefers wet mesic to dry soils in either full sun or partial shade. The red spherical rose hips (about 8 mm long) ripen for harvest from late August to late October. Several yellowish or reddish-brown seeds hairy on one side (4–5 mm long) can be recovered from the rose hips. Pasture rose has a Coefficient of Conservatism of 4 indicating its ease of restoration; however, it is a challenge to obtain seed germination. Several methods are recommended. A germination procedure recommended by Prairie Moon Nursery [11] requires seed scarification followed by 2-month cold-moist stratification; the seeds may germinate in the second growing season or later. According to Deno [12], several cycles of 3-month cool-moist (40°F) and warm-moist (70°F) in the dark have been tested with good results. Accordingly, one set of several germination cycles gave the following results: 40, 70°F (18% in 2–4 days), 40, 70°F (43% in 3–14 days), 40, 70°F (12% on the fourth day), and 40, 70°F (2%). Another set of germination cycles gave the following results: 70, 40, 70°F (60% in 2–8 days), 40, 70°F (26% on the second day), and 40, 70°F (1%). Sowing outdoors in the fall gave 38% germination in April through May [12].

2.107 **Rosa setigera* (Rosaceae Family), Prairie Rose or Climbing Rose

As one common name (climbing rose) indicates, *Rosa setigera*, prairie rose, can attain some stature (height, 2–5 m or more). Two varieties, v. *setigera* and v. *tomentosa,* share somewhat overlapping regions. The species ranges in the north from New Hampshire west to Ontario, Wisconsin, Iowa, and Nebraska. From the northern limit it is found in all states south and east. The uninitiated may mistake *R. setigera* for the aggressive alien multiflora rose, as both are bushy with climbing canes and armed with recurved prickles (occasional with *R. setigera*). However the much larger pink to whitish-pink flowers (4–8 cm broad, blooming in mid-July) distinguish *R. setigera* from multiflora rose, which has pyramidal clusters of abundant small white flowers (2–4 cm broad). As illustrated on page 114, the pinnately compound leaves of *R. setigera* are distinctive with only three leaflets (sometimes five), whereas the multiflora rose has seven or nine leaflets. This plant is found in wet-mesic to dry-mesic soils in full sun or partial shade. Its Coefficient of Conservatism is rated at 5. When the rose hips ripen to a red color, they can be collected. The hips (8–12 mm long) contain several yellowish seeds that are hairy on one side (3.8–4.2 mm long). For germination the seeds should be cool-moist stratified for 2 months [11], but germination may occur in a second growing season. Fall planting is a good option. This rose may not be evident for several years after planting.

2.108 **Rosa virginiana* v. *virginiana* (Rosaceae Family), Virginia Rose

Rosa virginiana v. *virginiana*, Virginia rose (height, 0.15–2 m), is found in parts of the Midwest (Ontario, Illinois, Missouri, and Arkansas), but it is more common eastward (illustration on page 114). In the East it is found from Newfoundland and Quebec south to Georgia, Tennessee, and Alabama, except its absence in South Carolina. One other variety, *lamprophylla*, is found only in parts of New England. The Virginia rose possesses pinnately compound leaves with 5–11, usually 5, leaflets. It is characterized by its strong thick-based needle-like prickles, which are either straight or hooked slightly downward and located on the stem immediately below the attachment with the leaf (infrastipular). Virginia rose requires wet-mesic to dry-mesic soils in full sun or partial shade. Pink flowers appear in mid- to late June. The developing fruit has spreading sepals, which soon drop from the red fruit. Seeds can be collected from the fruit anytime in the late fall. Because germination needs cool-moist stratification, fall planting is recommended. Germination may require a second growing season. Deno [12] found 80% seed germination (2–8 days) in the dark after cycles of 3 months each of warm-moist (70°F) and cool-moist (40°F), and then germination occurred after switching to warm-moist (70°F).

2.109 **Rudbeckia fulgida* (Asteraceae Family), Orange Coneflower

This attractive flower, *Rudbeckia fulgida*, orange coneflower (height, 0.4–1 m), is commonly cultivated in flower gardens (illustration on page 114). One clone named "Goldsturm" is a prized horticultural specimen known by some authorities as *Rudbeckia fulgida* v. *sullivantii* (e.g., [6]), but others classify it as *R. sullivantii* [13], or *R. speciosa* v. *sullivantii* [14, 16]. This confusion about its categorization as a species should be amenable to DNA analyses. With a Coefficient of Conservation of 7, *R. fulgida* is relatively difficult to establish; however, it is easy to cultivate in weeded beds. It prefers wet-mesic to dry-mesic soil in either full sun or partial shade. Because the plants definitely do not thrive in periods of drought, it is advisable to plant on the wet side of the recommended soil moisture. The seed-heads can be clipped from mid-September to mid-October. The dark gray seeds (3 mm long by 0.7 mm wide) are ridged longitudinally. Scraping over quarter-inch hardware cloth will loosen the seeds and partially purify by sieving. Seeds can be treated with 2 months of cool-moist stratification or placed in the soil without treatment [11]. In my hands fall planting resulted in little germination the next spring, but germination was very good the second growing season. Deno [12] quotes others that the optimum germination temperature is 82 to 90°F with some germination occurring at 70°F.

2.110 **Rudbeckia hirta* (Asteraceae Family), Black-Eyed Susan

Rudbeckia hirta, black-eyed Susan (height, 0.3–0.9 m), is described by various authorities as either sometimes annual, biennial, or perennial; however, it is agreed that the plant is largely biennial and sometimes blooms the first year if conditions are favorable (illustration on page 115). It is likely the plant dies after setting seeds. There are a total of four varieties of *R. hirta* with differing ranges. The most common varieties are v. *pulcherrima* and v. *hirta*. The variety *pulcherrima* is found in all the southern provinces of Canada and all the lower 48 states of the U.S., except Arizona and Nevada. Variety *hirta* is found mainly in the East and eastern Midwest with the exception of its absence in the far southern states. With a low Coefficient of Conservatism of 2, it definitely thrives best in disturbed soil, such as new restorations. Black-eyed Susan thrives in most soils ranging from wet-mesic to dry. It decreases dramatically as the prairie matures, but it persists at a low level of abundance. Thus, its value in restorations is to give a good show of yellow blooms before other restored species are evident. *R. hirta* blooms in early July, and seed-heads can be clipped in early to mid-August when the stems become brown. Seed-heads should be stored in paper bags for drying. Blackish quadrangular seeds (1.6–2 mm long by 0.3–0.4 mm wide) can be obtained by vigorous thrashing/shaking the dried seed-heads, or one can scrape the seed-heads over quarter-inch hardware cloth. Seed purification can be accomplished by sieving through window screen. For germination seeds require either no treatment or benefit from 2 months of cool-moist stratification [11]. Deno [12] found 90–100% germination with all treatments tested. At 70°F, germination occurred in 6–8 days.

2.111 **Rudbeckia subtomentosa* (Asteraceae Family), Sweet Black-Eyed Susan or Sweet Coneflower

Rudbeckia subtomentosa, sweet black-eyed Susan (height, 0.6–1.8 m), is a perennial that has a lifetime of two decades or more under optimum conditions (illustration on page 115). It prefers wet-mesic to mesic soils in prairies and savannas. In the East it has a sporadic occurrence in Massachusetts, Connecticut, New York, and North Carolina. I have been very successful growing this plant in Pennsylvania. In the northern Midwest its limit is Michigan, Wisconsin, Iowa, and Kansas. The southern limit in the Midwest is Texas, Louisiana, Alabama, and Tennessee. A unique characteristic of sweet black-eyed Susan regards the shape of the leaves, from three-parted, two-parted, or one-parted ovate to ovate-lanceolate. Crushing the leaves or seed-heads generates a sweet fragrance. Its Coefficient of Conservatism is rated at 5. Sweet black-eyed Susan is moderately easy to restore and responds very well to cultivation. A long flowering period occurs in early August to late September, which makes it a desirable addition to your restoration. The seed-heads are ready for clipping when the stems dry to a brown color from mid-September to early November. After storing in paper bags for drying, seeds can be thrashed from the seed head, but this process may prove to be more difficult than experienced with *R. hirta* described above. Scraping over quarter-inch hardware cloth may be required to obtain better yields. The blackish-brown seed (about 3 mm long) is somewhat more triangular than the seed of *R. hirta*. For germination, seeds require either a 2-month cool-moist stratification or no treatment [11]. In my hands, a late fall planting was successful.

2.112 **Ruellia humilis* (Acanthaceae Family), Wild Petunia or Hairy Ruellia

Ruellia humilis, wild petunia (height, 0.3–0.7 m), affords pale bluish-purple flowers about early July (illustration on page 115). Wild petunia prefers mesic to dry soils in either full sun or partial shade. The northern range of wild petunia is Pennsylvania west to Minnesota and Nebraska. From this northern limit it occurs in all states south and east, except New Jersey, Delaware, and South Carolina. Its Coefficient of Conservatism is a rather low 3. The flower corolla is short lived, dropping off to reveal the developing capsule. When mature, the smooth capsule is tan to brown (10–16 mm long). Seed collection is a challenge because the capsules shatter, sending the flat-circular seeds (3–4 mm in diameter) flying for some distance. The seeds have a small tail attached, which may enhance their aerodynamic ability. It is also difficult to pick individual capsules from the stems. One alternative is to clip the entire stem containing several capsules per stem; this should be timed in early to mid-October for maximum maturity of the capsules, but before most capsules have shattered (times may vary by region). The stems plus leaves and capsules should be loosely packed in paper bags, and the top of the bag fastened shut, such as with staples or paper clips. The bags should be placed out of the rain in full sun, which stimulates shattering of the capsules. While drying in the sunlight, snapping sounds occasionally can be heard within the bags. For germination [12] either fresh or 6-month dry-stored seeds are best germinated after a specific 14-week cool-moist (40°F) period before shifting to warm-moist (70°F). Exposing seeds to a cool-moist (40°F) temperature for various times of 0, 4, 6, 8, 10, 12, 14, 18, and 22 weeks before shifting to warm-moist (70°F) gave the respective percent germinations of 0, 2, 6, 27, 29, 47, 100, 75, and 88%. Fall planting is one option that gave results for this author.

2.113 **Salix humilis* (Salicaceae Family), Prairie Willow

Salix humilis, prairie willow (height, 0.5–3 m) is found as two varieties, v. *humilis* and v. *tristis*, sharing much of the same region in the East and Midwest. The variety *humilis* additionally occurs in Newfoundland and Labrador west to Manitoba. It is a shrubby willow that does not reach a great height. See illustration on page 115, showing the plant well after early spring flowering. Prairie willow grows in full sun or partial shade on mesic to dry soils, mostly on hillsides. Although its Coefficient of Conservatism is 5, this shrub is not at all common. Deer heavily browse prairie willow, to its detriment. Certain willow species will root from clippings, but most of my efforts have not succeeded with prairie willow clippings placed in wet sand, even when aided with a commercial root-stimulating hormone, Rootone F from Union Carbide. In another experiment, a branch cutting in the catkin stage was placed in water, and eventually shoots developed from the main branch. However, significant root development required more than 2 months. Subsequent planting of the rooted shoot was unsuccessful. The catkin stage is probably required for propagation from cuttings, but it may prove to be difficult. The catkins (1–3 cm long) emerge in April before the leaves. The capsules or fruit are pear shaped (4–8 mm long) and covered with fine short hairs.

2.114 **Schizachyrium scoparium* v. *scoparium*, *Andropogon scoparius* (Poaceae Family), Little Bluestem Grass

Of three varieties, the principal variety, *Schizachyrium scoparium* v. *scoparium*, little bluestem grass (height, 0.5–1.5 m) is generally found on mesic to dry prairies spread widely over the U.S., except Oregon and Nevada. See illustration on page 115. It is also found in the southern provinces of Canada. Little bluestem was introduced into Hawaii where it has become a nuisance weed in places, such as Volcano National Park. The other two varieties have a more restrictive distribution, mainly in the South. The orange-brown appearance of little bluestem during winter gives added incentive to include this species in a prairie. As might be expected, little bluestem grass has differentiated into several regional populations with different characteristics. In the 1940s, researchers collected seeds widely over the U.S. Subsequent cultivation of the grass in one location showed that a wide variance in growth characteristics indicated a division into "ecotypes" [5]. Northern types flowered earlier, produced less dry matter, and survived overwintering compared to southern collections. As mentioned above, little bluestem grass offers less competition to flowering species (forbs) compared to tallgrass. Although this shortgrass has a median Coefficient of Conservatism, it is often an indicator of old undisturbed prairies. This author has noted a general succession over the years of *Sorghastrum nutans*, followed by *Andropogon gerardii*, and finally *Schizachyrium scoparium* v. *scoparium* plus *A. gerardii*. Seed can be collected from late September to mid-November, specifically when a light "fluff" appears on the seed-head. Other than stripping by hand, a simple device comprising a comb attached to the bottom lip of a feed scoop facilitates collection. The fluff plus attached seed (about 7 mm long) can be planted without seed treatment [11]. According to precise experiments [12], seeds germinated 50% in 4–9 days at 70°F. Temperature cycling of 3 months at a moist 40–70°F gave 62% germination in 2–11 days. To initially establish little bluestem from being overwhelmed by weeds, Maurey Brucker of Peoria, Illinois (personal communication) advocates regular mowing a few inches high for a few years; however, one must be certain that the mower is not contaminated with the seeds of alien species.

**Andropogon virginicus* v. *virginicus*, broom sedge, is another species superficially similar in appearance to little bluestem (see herbarium on Springer internet site), and it is commonly found in eastern states as well as the Midwest. The comparative appearance of the seed-head and seeds is the easy method to determine the difference between these two species. Broom sedge has decidedly more fluffy seeds than little bluestem. Without seeds and further examination, the winter stand of grass in the illustration could, in fact, be either species.

2.115 *Senecio* see *Packera*

2.116 **Senna marilandica, Cassia marilandica* (Fabaceae Family), Maryland Senna

Senna marilandica, Maryland senna (height, 0.9–2 m), is a relatively large legume that has a Coefficient of Conservatism of 4 (illustration on page 115). Maryland senna prefers wet-mesic to dry-mesic soils in full sun or partial shade; however, it does best on more moist soils. It ranges from Massachusetts west to Wisconsin, Iowa, and Nebraska, and from this northern limit south to the border. The most notable feature of the plant is the large pinnate leaves. The seed-pods can be collected any time after early October. The pods persist into the winter. The dark brown seeds are oblong shaped (5 mm long) and can be recovered by breaking the black pods apart. Recommended seed treatment for germination is scarification, followed by 10 days of cool-moist stratification and inoculation with specific nitrifying bacteria [11]. This author has easily restored this plant using a late fall planting without any seed treatment.

A species similar in appearance and preferred ecosystem, **Senna hebecarpa*, wild senna (see Springer internet site for seeds/seed-head), differs mainly in its large Coefficient of Conservatism of 9. This species ranges somewhat north of *S. marilandica*. According to *Gray's Manual of Botany* [14], the segments of the seedpod of *S. hebecarpa* are as long as they are broad, whereas in *S. marilandica* the segments are much shorter than broad. Additionally, the seed-pod of *S. hebecarpa* is notably pubescent/hirsute.

2.117 **Silene regia* (Caryphyllaceae Family), Royal Catchfly

Silene regia, royal catchfly (height, 0.7–1.5 m), displays brilliant red flowers in mid-July (illustration on page 115). Royal catchfly prefers to grow in mesic to dry-mesic soil in full sun or partial shade. It is my experience that the plant does not thrive during drought. When stressed, the plant seems susceptible to fungal blight. Royal catchfly is a highly desirable species, not only for its color, but because it is a conservative species (Coefficient of Conservatism of 9). It is difficult to restore; however, it responds to cultivation as a row crop or weeded bed crop, but under cultivation the plant tends to lodge (fall over). Royal catchfly has a limited range in seven states in the Midwest centered in Missouri. It is also found in five states of the Southeast centered in northwest Georgia. The seed capsule (about 10–20 mm long) can be picked when it is senescent brown from early August to early September. The capsules should be held upright until placed in a bag, as the seeds tend to spill out of the open top. The capsules can be crushed to more efficiently recover the shiny brown seeds (1.8–2.2 mm long). The seeds are shaped like a one-half to three-quarter moon. For germination, seeds respond to a 2-month cool-moist stratification [11]. Fall sowing gives success in my hands. Deno [12] found that seeds germinate at 70°F only after 3 months, but he quotes a source who claimed 95% germination at 70°F (in 2–10 days) if the seeds were placed outdoors in January for 3 weeks.

The red flower color and other similar characteristics are reminiscent of a related *Silene* species, **Silene virginica*, fire pink (not shown). Although it is rather variably distributed in the East and Midwest, I have seen it only once on a bald in the Great Smokies National Park.

2.118 **Silene stellata* (Caryphyllaceae Family), Starry Campion

Silene stellata, starry campion (height, 0.4–1 m), adds a touch of white to prairies or savannas. It is shown on page 115 associated with the yellow flowers of *C. palmata*. A cluster of four leaves is whorled at the stem nodes, which is a distinguishing feature of this *Silene*. Starry campion is found from Massachusetts south to Georgia and west to North Dakota in the north and Texas in the south. Its soil preference is wet-mesic to dry-mesic, and the Coefficient of Conservatism is rated at 6. Similar to *Silene regia* (above), stressed starry campion develops a blight, possibly fungal infection. Flowering occurs in late June, and the seed-head can be collected when the spherical-oval capsules (about 0.5–1.3 cm long) are a senescent tan in late July to mid-August. The three-quarter moon-shaped seeds (dark purplish-brown, about 1–1.3 mm long) can be collected after crushing the capsules. The seeds germinate after a 2-month cool-moist stratification [11], or the seeds can be successfully sowed in the fall.

2.119 *Silphium integrifolium* (Asteraceae Family), Rosinweed

Silphium integrifolium, rosinweed (height, 0.6–1.5 m) acquires its common name from the rosin-like material that is exuded from its wounds. In the illustration on page 116, rosin droplets can be seen on the leaves. Related members of the *Silphium* genus, especially *S. laciniatum* (see below), are also prolific producers of rosin. Rosinweed comprises six varieties, many of which share the same area; the variety *integrifolium* has the widest distribution. The *integrifolium* variety and the others are not found in the East Coast states, including Ohio and West Virginia. Variety *integrifolium* has a northern limit in Ontario west to Minnesota and South Dakota. From the north it is found in all states south to its eastern limit of Indiana, Kentucky, Tennessee, and Mississippi. Rosinweed grows in wet-mesic to dry-mesic soil, mainly in prairies. As indicated by its Coefficient of Conservatism of 5, it is easy to restore. Its peak flowering time is mid-July, and the seed-heads senesce to brown from late August to mid-September. At this time, individual seed-heads can be plucked from the plant. Dried seed-heads can be scraped and sieved over quarter-inch hardware cloth. Much extraneous material is obtained with the seeds. The tan-gray seeds (0.8–1.1 cm long) resemble sunflower seeds in shape, but they additionally have side wings about 2 mm wide with a deep notch in the wings at one end. For germination the seeds require a 2-month cool-moist stratification [11]. Fall planting is a good option.

2.120 **Silphium laciniatum* v. *laciniatum* (Asteraceae Family), Compass Plant

Silphium laciniatum v. *laciniatum*, compass plant (height, 1–3.5 m) acquires its common name for the alignment of the leaves facing east and west; that is, the leaf edges point north and south. This alignment becomes more random when the plant is in the flowering stage. See illustration on page 116 that shows the sunlight in the west shining through the north-south-oriented leaves. The inset on the upper right shows the flower, which is borne on a long stalk. The typical variety, *laciniatum*, extends its northern range from New York west to Ontario, Minnesota, North Dakota, and Colorado. Except for its exclusion in a large portion of the Southeast and the Middle Atlantic states of Maryland, West Virginia, and New Jersey, it is found in all states south and east of its northern limit. One other variety, *robinsonii*, is restricted to the Mid-South. Similar to rosinweed (above), compass plant produces rosin when wounded. Its preferred habitat is wet-mesic to dry-mesic soils in full sun, but it is more often found in wet-mesic and mesic soils. In contrast to many other species, compass plant thrives well in thick stands of tallgrass. It is easy to restore when planted with *Andropogon gerardii* or *Sorghastrum nutans*, presumably because compass plant is tall and develops a taproot deeper than the roots of tallgrass species. Rodents apparently feed on the tops of these taproots over winter, but this long-lived plant recovers the next summer. This author has noted a lifetime of compass plant at least three decades. Its Coefficient of Conservatism is 5. The peak flowering of compass plant is early August. The seed-heads are brown senescent disks (about 3 cm across), resembling miniature sunflower heads. They can be picked from mid-September to late October. The disk seed-heads should be picked only when they show little resistance to plucking from the plant; resistance indicates later harvest times are appropriate. The seed-heads should be broken apart giving seeds and much extraneous material, such as fines and sepals. If desired, the fines and smaller seeds can be sieved through quarter-inch hardware cloth; the sepals and larger seeds are retained on top. It is a time-consuming procedure to separate the smaller seeds from fines using a window screen sieve. The gray-brown oval seeds (1.3–1.5 cm long) are reminiscent of sunflower seeds; however, they are surrounded with wings about 2 mm wide with a notch in the wings at one end. For germination the seeds require 2 months of cool-moist stratification [11], or the seeds can be fall planted. Germination during the second season of some seeds is often encountered. I have observed a hybrid of *S. laciniatum* with *S. integrifolium* originating from a seed of a *S. laciniatum* plant. The hybrid has features reminiscent of both parents. Hybrids of *S. laciniatum* and *S. terebinthinaceum* are fairly common. This latter hybrid emanates from seeds of *S. terebinthinaceum* that were cross-pollinated with *S. laciniatum*.

2.121 **Silphium perfoliatum* v. *perfoliatum* (Asteraceae Family), Cup Plant

Silphium perfoliatum v. *perfoliatum*, cup plant (height, 1–2.5 m) is easy to restore in the proper ecosystem, as indicated by its Coefficient of Conservatism of 4 (illustration on page 116). The variety *perfoliatum* is most common, ranging from Maine and Quebec west to Ontario and North Dakota. From the north it ranges south to Oklahoma and southeastward. It is absent in a few coastal states, principally South Carolina and Florida. One other variety, *connatum*, is found only in the Southeast from Maryland to North Carolina. The species name, *perfoliatum*, and the common name, cup plant, describe the structure of the upper leaves, which surround the square stem. These upper leaves serve as cups that catch rainwater and dew. The lower leaves do not surround the stem, but the lower leaves follow the margins of the petiole to its termination with the stem. Cup plant prefers wet-mesic to mesic soils in full sun or partial shade; however, the plants growing in mesic soil almost always are found in partial shade. The seeds are gathered when the disk is senescent brown, from late August to mid-September. Daily seed collection visits are required for maximum yield. The seeds are similar to the other *Silphium* species described above (9–11 mm long) with wings 1–2 mm wide with a notch in the wings at one end. Fines can be removed from seeds by sieving with a window screen. For germination, the seeds require a 2-month cool-moist stratification [11]. Fall planting is an optional method.

2.122 **Silphium terebinthinaceum* v. *terebinthinaceum* (Asteraceae Family), Prairie Dock

Silphium terebinthinaceum v. *terebinthinaceum*, prairie dock (height, 1.2–3 m), is distinctive in prairies by its huge heart-shaped basal leaves that sometimes measure 0.6 m long by 0.3 m wide. See illustration on page 116 showing large ovate/cordate leaves in the middle ground. The inset (upper right) shows the flower, which is borne on a long stalk of 8–10 ft. The variety *terebinthinaceum* is typically found in the Midwest and Ontario in the north, but it is also known in the Southeast south of

Maryland, with the exception of Louisiana and Florida. Another variety, *luciae-brauniae*, is restricted only to four states from Ohio and south to Mississippi. The long flower stalk is almost bare, possessing very small and widely spaced leaves. The species name, *terebinthinaceum*, means "like turpentine"; however, it exudes less resinous material from wounds than both *S. laciniatum* and *S. integrifolium*. Similar to compass plant, prairie dock has a large carrot-like taproot (8–9 ft) that exceeds the root depth of tallgrasses. It is likely that the deep taproot permits prairie dock to grow in a range of soil moistures from wet-mesic to dry-mesic. It does seem to thrive best in wet-mesic prairies. Deer sometimes graze on the basal leaves, but the plant recovers. Heavy and repeated grazing by deer can be hard on prairie dock, causing it to decrease; otherwise, it is a long-lived plant. Prairie dock can be cross-pollinated by compass plant (compass plant is shown in the foreground in the illustration on page 116) giving a hybrid with characteristics of both plants. For some reason, these hybrids are more common in degraded prairies. Although prairie dock's Coefficient of Conservatism is only 4, it is more difficult to restore than the other *Silphium* species described above. It probably survives disturbance by regeneration from its taproot. Prairie dock starts to flower in early August, and the senescent brown disk can be collected from early September to late October when the disk gives little resistance to picking. Because the seeds drop easily when ripe, seed collection must be done on a regular basis. Fine material can be eliminated with a sieve composed of window screen. Insects often sever the flower stem; thus, seed yields can be poor. Seeds are similar in appearance to *S. perfoliatum* seeds described above, except the notch at one end is less pronounced. For germination, the seeds require 2 months of cool-moist stratification [11], or can be planted in the fall. After fall planting, germination can occur during the second season.

2.123 **Sisyrinchium albidum* (Iridaceae Family), White Blue-Eyed Grass

The flowers of *Sisyrinchium albidum*, white blue-eyed grass (height, 0.15–0.45 m), appear in mid-May, and they can range in color from white to pale blue (illustration on page 116). Occasionally, fairly blue-flowered varieties can be observed (see color herbarium on Springer internet site). *S. albidum* is one of two most abundant species found in Illinois [22]. Its northern limit is Maine west to Ontario, Wisconsin, Missouri, and Texas. It can be found in most states south and east, except it apparently is absent in most of the northeastern states. Unlike *S. angustifolium* (see Sect. 2.124), *S. albidum* does not have multiple stem-like peduncles (flower stalks) emanating from the stem; that is, the flowers and the leafy bracts (spathes) terminate on one stem (sessile). Leaves are 1–3 mm wide, and the stem can be up to 4 mm wide. White blue-eyed grass will grow in full sun or partial shade in mesic to dry soils. Its Coefficient of Conservatism is 4. Several seeds are contained in somewhat spherical capsules measuring 2–3 mm in diameter. When ripe, the pale brown capsules can be stripped from the flower head about mid- to late June, and upon drying the capsules tend to split open giving spherical seeds 1 mm or less in diameter. Removal of extraneous material can be accomplished by passing the seeds through a window screen sieve. The seeds should be given a

2-month cool-moist stratification to promote germination, after which these small seeds should be scattered on top of the soil early in the spring [11]. In my hands, a short 1- to 2-month stratification failed seed germination, but seeds germinated during a second growing season. Fall planting is recommended [11].

2.124 **Sisyrinchium angustifolium*, *Sisyrinchium graminoides* (Iridaceae Family), Stout/Pointed Blue-Eyed Grass

The flowers of *Sisyrinchium angustifolium*, stout blue-eyed grass (height, 0.1–0.5 m), appear in mid-June, and they generally are pale blue to violet blue in color (illustration on page 116). In Illinois, *S. angustifolium* is one of the two most common species of blue-eyed grass [22], and it is fairly common in undisturbed areas of the East. Its known range is quite large, from the northern limit of Labrador west to Ontario, Minnesota, and Kansas, and south and east from the northern extent to all states. This species is mainly characterized by broadly winged stems (3–5 mm wide), and one to three stem-like peduncles (flower stalks) about 2–15 cm long emanating from the top of the stem, which is enclosed by a large leafy bract. The flowers emerge from leafy bracts (spathes) at the top of the peduncles. These features are best observed as either gray-scale or color herbarium on Springer website. The preferred ecosystem of stout blue-eyed grass is either full sun or partial shade growing on wet-mesic to dry-mesic soil; however, it does best in more moist situations. Its Coefficient of Conservatism is 5. The somewhat spherical seed capsules (4–6 mm in diameter) ripen to a dark brown/blackish color in mid-July, at which time they can be stripped from the plant. Upon drying, the capsules split, releasing black spherical seeds (about 1–1.4 mm in diameter, microscopically pitted). According to one source, seeds are successfully germinated after a 2-month cool-moist stratification followed by scattering the seeds on top of the soil [11]. Another source [12] showed that cool-moist treatment was detrimental to germination, but there was an absolute requirement for light. Light-induced seed germination was 40% at 70°F in 6–9 weeks, whereas a 3-month cool-moist (40°F) period followed by 70°F was less effective, giving 15% germination.

2.125 *Solidago* also see *Oligoneuron* and *Euthamia*

2.126 **Solidago canadensis* (Asteraceae Family), Canada Goldenrod

Solidago canadensis, Canada goldenrod (height, 0.3–1.5 m), is a native weedy species (Coefficient of Conservatism, 1). As a species, it is found in all but one northern Canadian province and all of the U.S., except five southeastern states. There are five recognized varieties of *S. canadensis* [15]. The variety shown here is probably *S. canadensis* v. *canadensis* (illustration on page 117), but two other varieties, *gilvocanescens* and *hargeri*, are found largely in the same area. The variety *canadensis* is located in eastern to mid-Canada, and in an area one-third of the U.S. from the Northeast to the Midwest. It is nearly impossible to avoid this plant in prairie restorations. Some authorities recognize *Solidago altissima* as a separate species [13, 14], but another authority considers it as a variety, *Solidago canadensis* v. *scabra* [15]. *Solidago altissima* was introduced into Europe from America about 300 years ago. Consequently, it spread widely over Europe. Swiss workers [23] showed that in this short time *S. altissima* had differentiated into population phenotypes characteristic of the latitude from which it was collected. Plants from the northern populations were smaller and flowered earlier than the southern populations when grown in the same plot. This observation gives credence to the idea that seed collected from the same area should be used in restorations. Also, it is remarkable that differences appeared in only 300 years, arguing for epigenetics triggered by a genomic environmental switch, rather than caused by an actual mutational change. *Solidago canadensis* thrives in wet-mesic to dry-mesic soils in full sun or partial shade; it especially favors the more moist soils. Blooms begin to appear in mid-September; seeds are ripe when the pappus fluffs out in late October to mid November. This rhizomatous plant can be initially very aggressive in new restorations, forming extensive monoclonal patches, but it will sometimes decrease after several years. One can high mow or cut the plants back to about 0.15-m height during the very initial phase of flowering. In my hands, persistent cutting resulted in eventual replacement with mainly *Eryngium yuccifolium*, a plant that was largely hidden by the goldenrod canopy. Even though Canada goldenrod is an authentic prairie species and has certain attractiveness in flower, it is not particularly recommended for restorations. If you wish to plant this species, one can dry store the seeds for 6 months at either 40 or 70°F followed by germination at 70°F (during the second week); all other treatments were ineffective [12].

2.127 **Solidago graminifolia*, see *Euthamia***

2.128 **Solidago juncea* (Asteraceae Family), Early Goldenrod**

Although many species of goldenrod are found on the prairie, *Solidago juncea*, early goldenrod (height, 0.3–1.2 m) is uniquely an early bloomer in mid-August (illustration on page 117). Large serrated lanceolate or oval-lanceolate basal leaves are a distinguishing feature of this goldenrod. Early goldenrod's northern range extends from Maine and Quebec west to Manitoba and Minnesota. It is found in all states south and east of the northern boundary, except Iowa. The Coefficient of

Conservatism is 4. The preferred habitat is mesic to dry open areas, but this goldenrod also tolerates partial shade. Seeds can be collected when the pappus fluffs out from mid- to late September. The seed is 1.5–1.7 mm long with a pappus 2.6–3.3 mm long. Seed should be sowed on top of the soil after 2 months of cool-moist stratification [11], or alternatively sowed in the fall.

2.129 *Solidago rigida* see *Oligoneuron rigidum*

2.130 **Solidago speciosa* (Asteraceae Family), Showy Goldenrod

Solidago speciosa, showy goldenrod (height, 0.6–2 m), is truly a showy yellow plume when in flower in early October. See illustration on page 117. The flowering wands borne on side stems spiral up the main stem, which contribute to its showy nature. The reddish-brown stem of this goldenrod is also unique. Showy goldenrod is found in full sun or partial shade on mesic to dry soil. There are four varieties, of which three share a large portion of their range. The typical variety *speciosa* is found in the East and Midwest, excluding Florida and Alabama. Variety *rigidiuscula* is found in the Midwest and Great Plains states, whereas v. *jejunifolia* is restricted to the upper Midwest. For a goldenrod, its Coefficient of Conservatism is a fairly high at 7. It is difficult to restore in certain ecosystems but responds well in others. Showy goldenrod is easy to grow as a cultivated row crop. The seed-heads can be clipped when they appear fluffy from mid-October to mid-November, and then stored for drying in paper bags. Seeds readily blow away in the wind; thus, regular harvests are advised. After drying, the seeds plus pappus are easily removed by thrashing the seed-head or scraping over quarter-inch hardware cloth. The tan seeds (1.2–1.5 mm long by 0.2–0.3 mm wide) are topped by a fluffy pappus (4–5 mm long). The recommended germination procedure for seeds is cool-moist stratification for 2 months followed by early spring sowing on top of the soil [11]. Also, a late fall sowing on top of the soil was found to be satisfactory.

2.131 **Sorghastrum nutans* (Poaceae Family), Indian Grass

Sorghastrum nutans, Indian grass (height, 0.8–2.5 m), is a pioneer tallgrass that can progress across a degraded area in advance of other species; however, it slowly gives way to other grass species, such as *Andropogon gerardii* and *Schizachyrium scoparium*. See illustration on page 117. Rated at a Coefficient of Conservatism of

4, it is a successful species in new restorations. Indian grass prefers to grow in full sun or partial shade on mesic to dry soil. It is located throughout the U.S. and southern Canada, except for the far West. The flowering spikes appear in early September, and the seeds are ripe for hand stripping from the seed-head from late September to late October. Seeds should be stripped only when they offer little resistance of removal. The seeds should be spread out on a dry surface or screens to dry. The reddish-brown seed or grain (6–8 mm long) is flanked by hairy projections emanating from the base. Seeds require no treatment for germination [11]. Fall sowing gave good results.

2.132 **Spiranthes cernua* (Orchidaceae Family), Nodding Ladies' Tresses

Spiranthes cernua, nodding ladies' tresses (height, 0.15–0.6 m), is in the orchid family (illustration on page 117). Its orchid-like white flowers twist around the fleshy main stem in early September. This species prefers full sun or partial shade in wet-mesic to dry-mesic soil. Although this plant is not common, it has a fairly wide distribution from a northern limit of Maine and Quebec west to Ontario Minnesota, and Kansas, and from this limit it is located south and east everywhere, except Florida. It has a relatively low Coefficient of Conservatism of 4. Seed-heads can be collected in mid-October. The tiny seeds are abundant and disperse easily. Seeds should be scattered soon after collection on top of the soil. I have found this plant on electric power right-of-ways, indicating tolerance for disturbance. However, it is best to sow the seeds in established ecosystems devoid of weedy growth. The seeds reportedly contain two or more embryos.

2.133 **Sporobolus heterolepis* (Poaceae Family), Northern Dropseed

Sporobolus heterolepis, northern dropseed (height, 0.4–1 m), is truly the king of grasses (illustration on page 117). Its fine leaves fountain over in individual clumps or tufts that make it an ideal landscaping grass. With a Coefficient of Conservatism of 9, it is among the most conservative of grasses. It is often used as a marker of undisturbed prairies. Northern dropseed grass prefers either full sun or partial shade in wet-mesic to dry soil. Its northern range is Massachusetts, New York, and Quebec west to Saskatchewan and Montana. It is found south and east of the northern limit, except the Mid-South, much of the Southeast, upper New England, West Virginia, and New Jersey. As the frequency of flowering is dependent on conditions, flowering can be sporadic in early July to August. Seeds should be stripped from the seed-head when they are tan, not green, from mid-August to early October.

Because the seeds will drop shortly after ripening, harvesting must be done periodically for more than a month. The tan seed or grain is spherical (1.4–2 mm in diameter), and it requires no treatment for germination [11]. Provided that the area is relatively free of weed seeds, restoration is easy. Alternatively, the planted area can be weeded regularly, avoiding the tiny fine-leaved grass (the first year of growth is excruciatingly slow). Periodic treatment of an area with glyphosate for a full year before restoration can help with weed problems. Subsoil from construction sites was found to be ideal as it lacks weed seeds.

2.134 *Stipa* see *Hesperostipa*

2.135 **Symphyotrichum ericoides* v. *ericoides, Aster ericoides* (Asteraceae Family), Heath Aster

Symphyotrichum ericoides v. *ericoides*, heath aster (height, 0.3–1 m), belies its name by its heath-like aspect and density of small leaves (illustration on page 117). According to the USDA website (http://plants.usda.gov/java/nameSearch), there is some confusion about the identity of *Symphyotrichum ericoides* v. *ericoides* (also known as *Aster ericoides*). On the website two illustrations are given from Britton and Brown [13], one for *A. ericoides* and another for *Aster multiflorus*. It is my opinion that the *A. multiflorus* illustration appears to be correct, and the *A. ericoides* illustration was misidentified by Britton and Brown [13]. In early October heath aster forms dense clusters of tiny white flowers. It is fairly easy to propagate, and it responds well to row cropping for seed production. Such cultivated plants seem to have a lifetime of more than a decade and appear to spread by rhizome. It prefers full sun in mesic to dry soil. The variety *ericoides* has an extensive range from its northern limit of Maine and Quebec west to Manitoba, the Dakotas, Colorado, and Arizona. This aster has been documented in all states south and east of this northern range, except Kentucky, the Carolinas, Louisiana, and Florida. A Coefficient of Conservatism of 4 shows that heath aster does well in newly restored prairies. The flower head can be harvested in mid-November when fluff begins to appear. For recovery of seed, see procedure described below in Sect. 2.138. The seed plus fluffy pappus (3–4 mm long) requires no seed treatment for germination, but the small seeds should be sowed on top of the soil. According to Deno [12], satisfactory germination occurred at 70°F using either fresh seeds or seeds dry stored for 6 months at 40°F. In my hands, fall planting gave results.

**Symphyotrichum/Aster* × *amethystinus*, amethyst aster (height, 0.6–1.5 m; see herbarium on Springer internet site), is an interesting recurring hybrid of *Symphyotrichum ericoides* and *Symphyotrichum novae-angliae* [14, 15], and indeed, all three are sometimes found together in prairies. Generally, this hybrid has blended characteristics of both parents. The amethyst flowers appear in early October. Its habitat is moist to mesic full-sun prairies or savannas. Although this aster is not encountered often, it has a mid-ranking in its Coefficient of Conservatism.

2.136 **Symphyotrichum laeve* v. *laeve, Aster laevis* (Asteraceae Family), Smooth Aster

In contrast to most asters, *Symphyotrichum laeve*, smooth aster (height, 0.5–1 m), is smooth (glabrous) throughout as the name implies (illustration on page 117). The leaves additionally have a whitish bloom (glaucous). This aster furnishes attractive blue flowers in mid-September. Smooth aster prefers wet-mesic to dry-mesic soils in full sun or partial shade. The typical variety *laeve* has a northern limit of New Brunswick west to Manitoba. It is found in all U.S. states south and east of the northern area, except the Dakotas, Texas, and Florida. As it is difficult to establish on many soils, an assigned Coefficient of Conservatism of 8 seems appropriate. Smooth aster thrives on black prairie silt-loam. The seed-head can be collected when fluffy from late September to early November. The seed ripens at various times and disperses easily, especially in windy weather, so multiple and selective collections are advised. Seed is harvested from the seed-head as described below for *Symphyotrichum oolentangiense*. The seed plus fluffy pappus is about 6–7 mm long. No specific seed treatment is required for germination [11], but fall planting is somewhat successful.

2.137 **Symphyotrichum novae-angliae, Aster novae-angliae* (Asteraceae Family), New England Aster

When *Symphyotrichum novae-angliae*, New England aster (height, 0.6–2 m), flowers in early October, it compensates for its homely nondescript appearance of the nonflowering plant (see illustration on page 118). Very few of the prairie asters can compete with the brilliant purple and relatively large bloom of this species. A variety with a rose-colored inflorescence also can be occasionally observed. New England aster has a range far more extensive than its namesake, mainly in areas not too hot or dry. In the north it is found from New Brunswick west to Manitoba, North Dakota, Montana, and British Columbia/Washington. From Washington it is found in all states south and east, except Idaho, Nevada, Arizona, Texas, Louisiana, and Florida. With a Coefficient of Conservatism of 4, it enjoys wide distribution in tolerating disturbance. This aster prefers full sun with wet- to dry-mesic soils, and occasionally can be initially aggressive in newly restored prairies. It thrives better under wet-mesic conditions. This author estimates its lifetime at somewhere between 5 and 10 years. Seed-heads can be collected in early November when reddish-brown fluff appears. See below in the Sect. 2.138 for description of seed collection from the seed-head. The seed plus attached pappus measures about 5–6 mm. Prairie Moon Nursery [11] recommends a 2-month cool-moist stratification for germination; however, it was shown by Deno [12] that seeds germinate at 70°F (95–100%) in 3–8 days from either fresh seed or seed dry stored 6 months at 70 or 40°F. Fall planting also gives good spring germination.

2.138 ******Symphyotrichum oolentangiense* **v.** *oolentangiense, Aster azureus* (Asteraceae Family), Azure Aster

Symphyotrichum oolentangiense v. *oolentangiense*, azure aster (height, 0.3–1.2 m), is desirable for adding bright azure color during early October (illustration on page 118). This typical variety of azure aster has its northern limit from New York west to Ontario, Minnesota, and South Dakota. It is found in all states south and east of South Dakota, except New England, South Carolina, and the Mid Atlantic states south to Virginia and Kentucky. It is very difficult to maintain dense populations, as its Coefficient of Conservatism is fairly high at 7. It is more likely to find this plant scattered widely in its preferred habitat of full sun or partial shade on mesic to dry soils. This author has successfully cultivated this aster by row crops or in beds, but after a few years of natural die off and seed migration by wind, there is a resistance to intensive cultivation in one location. It is this author's estimate that the plant's lifetime is 2–6 years. The flower heads can be clipped when the fluffy pappus appears in late October to mid November. When the flower head begins to develop "fluff," the top of the plant can be harvested with clippers for future seed collection. Harvested tops should be packed loosely in paper bags for drying in air. When dry, the fluffy seed can be shaken loose. It is convenient to scrape the sample on top of quarter-inch hardware cloth, which separates the stems from the seed. Many commercial suppliers separate the fluff (pappus) from the seed, but this is unnecessary for use in restoration. The seed plus pappus is about 5 mm long. The seeds require no treatment for germination [11]; however, fall planting has proved to be successful in my hands.

2.139 ******Symphyotrichum pilosum* **v.** *pilosum, Aster pilosus* (Asteraceae Family), Frost or Hairy Aster

Symphyotrichum pilosum v. *pilosum*, frost or hairy aster (height, 0.3–1.2 m), tends to be a very weedy species that is a common aggressive invader of newly restored prairies (Coefficient of Conservatism, 0). See illustration on page 118. The northern range of the typical variety *pilosum* extends from New Brunswick west to Ontario, Minnesota, and South Dakota. It is known in virtually every state south and east of South Dakota. This species tends to be highly variable, but one other variety, *pringlei*, is recognized. Because this fall-blooming species (late September to mid-October) tends to disappear as the prairie matures, the best advice is to tolerate and enjoy this native white aster while present. Frost aster prefers full sun in wet-mesic to dry soils. It is not advisable to purposefully restore this species unless diversity is desired. To propagate, the seed can be collected in late November to early December, and the seed can be fall planted if one is willing to tolerate this early-succession aggressor. For germination, Prairie Moon Nursery [11] recommends a 2-month cool-moist stratification followed by sowing on top of the soil.

2.140 **Teucrium canadense* v. *canadense* (Lamiaceae Family), American Germander or Wood Sage

Teucrium canadense v. *canadense*, American germander (height, 0.2–1 m), is a member of the mint family (illustration on page 118). Germander prefers full sun or partial shade on wet-mesic to mesic soil, but it will grow on dry-mesic soil if kept in partial shade. As a species, germander occupies all the U.S. and most of the lower provinces of Canada; it is very common. There are three varieties, of which v. *canadense* and v. *occidentale* are most likely to be encountered in the tallgrass area. The typical variety, *canadense*, is found in Canada from New Brunswick west to Ontario, and in the US it is located from the East Coast to the Great Plains, including New Mexico and Arizona. With a Coefficient of Conservatism of only 3, it is easy to restore in its proper ecosystem, and it spreads by rhizomes. This mint enjoys wide distribution over the Midwest and East. It does decrease with time in late succession prairies. The flowers appear in early to mid-June, and the seed-heads are ready for collection from early September to mid-October. After the seed-heads dry, the seeds can be rubbed loose from the receptacle (calyx). The reddish-brown seeds (1.5–2.4 mm long) are somewhat half-spherical and wrinkled. Seeds will germinate after a 2-month cool-moist stratification [11]. Fall planting is an option.

2.141 **Thalictrum dasycarpum* (Ranunculaceae Family), Purple Meadow Rue

Thalictrum dasycarpum, purple meadow rue (height, 0.6–2 m), can be found in partial shade or full sun in wet-mesic to mesic soils (illustration on page 118). Its northern range extends from Quebec west to British Columbia and Yukon. In the U.S., it is found in all areas with the exceptions of New England, the Southeast, Oregon, Nevada, and California. Purple meadow rue has an interesting overall aspect, which includes the unusual leaf pattern and height of the inflorescence. It has a medium Coefficient of Conservatism of 5. It blooms in June, and sets seeds in mid-August. The seeds are dark and ribbed longitudinally (3.8–5.5 mm long). When the seeds turn from green to a dark color, the seed-head can be clipped, and seeds can be easily rubbed or scraped free using quarter-inch hardware cloth sieve.

For proper germination the seeds should be sowed in late fall. Alternatively, the seeds can be stratified under cool-moist conditions for 2 months, followed by sowing in the early spring [11].

2.142 ******Tradescantia ohiensis*** (Commelinaceae Family), Ohio Spiderwort**

Tradescantia ohiensis, Ohio spiderwort (height, 0.3–0.9 m), blooms in late May with predominantly bright blue flowers, but purple, rose, and white forms are occasionally found (illustration on page 118). Its habitat is diverse, from full sun to partial shade in wet-mesic to dry soils, with nearly all types of soil structure. It has a wide geographical distribution from its northern limit of Maine and Ontario to Minnesota and Nebraska. Its southern limit is all states south and east of Nebraska. Perhaps by consequence of its success, its Coefficient of Conservatism is a low 3. The capsules containing the fruit and seed can be collected from early to late July. As most members of the *Tradescantia* genus, the seed capsules shatter readily when dark brown; thus, it is advisable to pick the capsules when beginning to senesce to yellow green. The oblong seed is 2–4 mm long and pitted. Seeds should be planted in late fall. Alternatively, it is suggested that seeds should be given a 4-month cool-moist stratification followed by planting in early spring [11]. A fall sowing of seed in weed-free/weeded beds flowered the first growing season about 1 month later than the usual blooming time.

2.143 ******Tridens flavus*** v. ***flavus, Triodia flava*** (Poaceae Family), Purpletop Grass**

Tridens flavus v. *flavus*, purpletop grass (height, 1–1.5 m), is otherwise known as tall redtop, which in name can be confused with redtop, *Agrostis alba*. *Agrostis alba* is a similar-appearing grass from Eurasia. The Eurasian grass has a seed-head that is reddish with much more branching, whereas, comparatively *Tridens flavus* has a dark purple seed-head with fewer branches that droop more (illustration on page 118). Flowering occurs in late summer. Purpletop prefers to grow in full sun on gravelly/sandy hillsides in mesic to dry soil. However, mesic soil often proves to give purpletop too much competition from other plants. Lush stands have been observed on sands east of the Illinois River and on hillsides composed of shale soil. The typical variety *flavus* can be found in every state south and east of its northern limit of New Hampshire west to Ontario, Minnesota, Nebraska, and New Mexico. It also occurs in California. One other variety, *chapmanii*, is only found in the Southeast and Mid-South. Purpletop has a Coefficient of Conservatism of only 1, indicating its ability to tolerate disturbance. In mid- to late October seeds can be stripped from the seed-head, preferably holding a 5-gallon bucket below to catch errant seeds. Seeds should be sowed in late fall or given a few months of cool-moist treatment [11]. It has been my experience that either a spring burn or an early summer mowing promotes the growth of purpletop, which gives it an advantage over cool season grasses.

2.144 **Tripsacum dactyloides* (Poaceae Family), Eastern Gama Grass

Tripsacum dactyloides, eastern gama grass (height, 1.2–2.5 m), is a wide-bladed grass of open bottomlands (wet-mesic to mesic) bearing a rather large seed and seed-head (illustration on page 118). The blades of grass have rough edges that easily cut the skin. This grass is found in all states south and east of its northern limit, ranging from Massachusetts and New York west to Michigan, Illinois, Iowa, and Nebraska. With a Coefficient of Conservatism of only 4, it is easy to establish in the preferred ecosystem. It not only spreads by rhizome, but its overbranching seed-heads are efficient in seeding adjacent areas; that is, it can be aggressive. Intense prairie fires can damage the large clumps of rhizomes in older plants from which it can slowly recover. Eastern gama grass flowers in early July, and it continues to flower for a month or two. Several seeds are stacked on each of the multiple seed-heads (usually 2–3) and ripen to a tan or brown color from the top down. At the time of ripening they can be easily broken off with little resistance. The unripe green seeds resist breakage. Because ripe seeds readily drop off the seed-head and continue to ripen from early August to early October, many biweekly or weekly harvest sessions are required. Tan cylindrical seeds are rather large (0.7–1 cm long by 0.5–0.6 cm wide). The seeds can be fall planted or stratified with cool-moist conditions for 2 months [11]. Seeds may germinate after 2 years.

2.145 **Vernonia fasciculata* ssp. *fasciculata* (Asteraceae Family), Western or Common Ironweed

Vernonia fasciculata ssp. *fasciculata*, western ironweed (height, 0.6–1.8 m), lends a bright purple to the prairie or savanna. See illustration on page 118. Except for its presence in New York and Massachusetts, western ironweed is localized in the Midwest. Its northern limit is Ohio west to Minnesota, Manitoba, and South Dakota. The southern limit is from Kentucky west to Mississippi, Arkansas, and Oklahoma. One other subspecies, *corymbosa*, has a somewhat more western range. With a Coefficient of Conservatism of 5, its principal habitat is wet-mesic to mesic soil. The seeds persist on the seed-head quite late in the fall, and can be identified by the rusty-tan to purplish-tan color of the fluffy pappus. Seed-heads can be collected from mid-September to mid-October. The pappus displays a "paintbrush" appearance for a month or more, at which time the seeds are not ready for harvest.

When the pappus fluffs out in a more circular pattern, the seeds are ripe. The seed is about 3–3.5 mm long and 0.5 mm wide where it attaches to the pappus. The pappus is about 6 mm long. For germination seeds can be treated with a 2-month cool-moist stratification [11]. Fall planting gives results.

There are several related species in the tallgrass prairie area, particularly **Vernonia gigantea/altissima*, *V. missurica*, and *V. baldwinii* (not shown). According to Denison [24], the floral bracts of *V. fasciculata* are upright and rounded (blunt), edged with fine hairs on the margins (see herbarium at Springer website). In the East **V. noveboracensis* (New York ironweed) is common; it is most easily identified by purple fibers ascending from the tips of the floral bracts.

2.146 ****Veronicastrum virginicum*** (Scrophulariaceae Family), Culver's Root

Veronicastrum virginicum, Culver's root (height, 0.8–2 m), is notable by its whorl of three to seven leaves and one or more terminal spikes of mainly white flowers (illustration on page 119). It is usually observed in wet-mesic to mesic bottoms in prairies, savannas, and occasionally woodlands. Culver's root's northern limit is Maine and Ontario west to Manitoba, and in the U.S. its habitat is all states south and east of North Dakota. Its Coefficient of Conservatism is 6. Deer have been known to browse on the plant. Culver's root blooms in early July, and the seed harvest can occur after the spikes turn brownish black from late September to late October. At this time one can simply strip off the capsules (2.5–4.5 mm long) by hand, but this is "quick and dirty" because capsules are mixed with the minute light-brown seeds (0.5–0.7 mm long), which are slightly elliptical and circular in cross section. (See Springer website for images of seed capsules and seeds.) Seed purification is an unnecessary effort for those not interested in the sale of seed. Seeds do not need treatment for germination, but these very small seeds should be sowed on top of the soil. For germination Deno [12] showed a light requirement and that a cool-moist treatment was unnecessary. Accordingly, 70°F in the light gave 62% germination in 6–12 days, but 0% at 70°F in the dark. Seeds placed outdoors in February germinated 33% in April. Dry storage of seeds for 2 years resulted in seed death.

2.147 ****Viola sororia***, ****Viola cucullata***, or ****Viola affinis***, ***Viola papilionacea/pratincola*** (Violaceae Family), Common Blue Violet

With a Coefficient of Conservatism of only 1, common blue violet (height, 10–20 cm) will volunteer, that is, if any seed source is nearby. Three species (*Viola sororia*, *V. cucullata*, and *V. affinis*), and especially their hybrids make "common blue violet" difficult to identify. They have a wide distribution in the East and Midwest, as well as southern Canada from the eastern to the mid-western provinces. According to Keener and Leitzel, as well as Russell [25], *V. papilionacea/ pratincola* can be referable to other species, and has been used as a "catch-all for almost any glabrous or near-glabrous stem-less blue violet with uncut leaves." *V. cucullata* can be identified by the flower stems being longer than the leaf stems, by the lateral petals bearded with short club-shaped hairs, and by sepals long and tapering, acute at the tip (habit, open wet areas). The illustration on page 119 is probably *V. cucullata.* Both *V. sororia* and *V. affinis* have flower stems more or less equal to the leaf stem. *V. affinis* has lateral petals bearded with long thread-like hairs, spurred petals densely bearded, and sepal auricles (habit, wet openings or woodlands). *V. sororia* has spurred petals glabrous or a few hairs within, and leaves moderately to densely long-pubescent throughout (habit, meadows, open woods, disturbed areas). The deep violet flowers appear in early May, as well as in the fall. These violets can be found in prairies, savannas, or woodlands on wet-mesic to dry-mesic soil; however, on drier soils these species prefer more shade. Common blue violet hybridizes with numerous other *Viola* species [14]. The cross with *Viola pedatifida* (*Viola* × *bernardi*) is the most common hybrid seen in prairies. In late May to early June the elliptical seed capsules (yellowish- to purplish-brown, 10–14 mm long) can be collected and dried. Seeds obtained from split capsules are purplish brown, brown, to olive black depending on the species. Generally, the seeds are 1.8–2.2 mm long. Seeds can be cool-moist stratified for 2 months or sowed in the late fall on top of the soil [11]; however, germination behavior is more complex, as shown by several germination experiments completed by Deno [12]. He found that the best germination of 68% in March was obtained by alternating 3-month temperature fluctuations of 70, 40, and 70°F, then placing the seeds outdoors in January. Germination of 54% in March occurred simply by collecting the seeds in June and placing them outdoors. There was a stimulation by light as shown by a 38% germination by moist incubation 3 months each with 70, 40, and 70°F, whereas the corresponding dark treatment gave no germination. Seeds dry stored for only 6 months at 70°F were not viable. Because evolution dictates survival of the fittest, the natural seed shatter in May/June is the best strategy for propagation.

2.148 *Viola pedatifida* (Violaceae Family), Prairie Violet

Viola pedatifida, prairie violet (height, 8–18 cm), has a resemblance to *Viola pedata* (see Chap. 3), but it can be identified more thoroughly by the differences described in the *V. pedata* description. See illustration on page 119. The leaves are palmately three-parted and each segment is again three-cleft, these being cleft further into two to four lobes. Prairie violet is limited to the Midwest and Great Plains in Canada and the U.S. In the north it is located from Ohio and Ontario west to Montana and Alberta, and in the south it can be found from Arkansas west to Oklahoma, New Mexico, and Arizona. Violet to reddish-violet blooms appear in early May. As described above, prairie violet occasionally will cross with common violet, giving a hybrid leaf structure intermediate in appearance. The preferred habitat of *Viola pedatifida* is full sun or partial shade on mesic to dry-mesic soil. It has a high Coefficient of Conservatism of 9. Seeds can be collected in June from capsules (8–12 mm long) when they just turn yellow green. However, higher seed yields usually are obtained in the fall from cleistogamous flowers (small self-pollinating flowers that remain closed). These capsules are harvested from mid-August to late October after turning to yellowish-tan color. Because the capsules split open to release the seeds, regular visits to harvest the capsules are necessary. The somewhat egg-shaped seeds are light brown and 1.7–2 mm long. Studies [12] revealed that 78% germination was achieved in April when seeds were placed outdoors in October. Alternating temperatures in the presence of moisture (all in the dark) of 70, 40, and 70°F of 3-month duration gave reasonable results of 30% germination on the sixth day; there was no light effect.

2.149 **Zizia aptera* (Apiaceae Family), Heart-Leaved Golden Alexanders

Zizia aptera, heart-leaved golden alexanders (height, 0.3–0.7 m), is a rather rare *Zizia* species that can be distinguished from the more common *Zizia aurea* by its heart-shaped basal leaves (illustration on page 119). However, it has a fairly wide distribution over much of the U.S. and southern provinces of Canada. It is absent from most of the Southwest, the upper New England states, Louisiana, Nebraska, and Kansas. The leaves on the flowering stem are largely similar to *Z. aurea*. The Coefficient of Conservatism of heart-leaved golden alexanders is rated at a rather high 9; thus, it is difficult to restore. *Zizia aptera* does respond well to row or bed cultivation. Its preferred habitat is full sun or partial shade on mesic to dry-mesic soil. The flowers appear in mid-May, and the seed-head can be clipped when it becomes senescent brown from early August to early September. Seeds are easily removed by manually scraping the seed heads over quarter-inch hardware cloth. The oval brown seeds are about 3 mm long by 2 mm wide. This species is a cool season plant that requires sowing either in the late fall or early spring; early spring plantings should be preceded with a 4-month cool-moist stratification of the seeds [11]. However, Deno [12] showed that 44% germination could be achieved in 2–6 weeks simply by warm-moist (70°F) treatment in the light, but no germination occurred in the dark. Percent germination dropped to 10% after dry storage for 6 months. In my hands a late fall sowing on top of the soil with a light cover of sand or glyphosate-killed thatch produced good results.

2.150 **Zizia aurea* (Apiaceae Family), Golden Alexanders

Zizia aurea, golden alexanders (height, 0.3–1 m), is often seen in damp prairie areas as a carpet of yellow by mid-May (illustration on page 119). It is widely distributed from Quebec to Manitoba in Canada, from Maine to North Dakota in the northern

U.S., and south to the U.S. border. Its preferred habitat is wet-mesic to mesic soils in full sun or partial shade. In dry-mesic soil it is largely confined to more shaded areas. Its Coefficient of Conservatism is 6. In appearance *Zizia aurea* slightly resembles the alien weed *Pastinaca sativa*, wild parsnip; this can make it difficult to spot wild parsnip for eradication. The wild parsnip flower tends to be more erect on one stem, whereas *Z. aurea* is a more sprawling plant. The seed-heads can be clipped when they are senescent brown in late July to mid-August. Seeds can be removed by manually scraping the seed-heads over quarter-inch hardware cloth. The oblong brown seeds are 3–4 mm long by 1.5–2 mm wide. For seed germination, this author simply sows seeds on top of the soil in the fall. Deno [12] has shown that gibberellic acid-3 strongly promotes germination at 70°F in the dark (100% within 15 days), and seeds dry stored for 6 months at 70°F germinated 60% at 70°F in the presence of gibberellic acid-3. Deno (personal communication) postulated that in the soil fungi may provide the gibberellic acid necessary for germination.

Mesic Soil and Adaptive Species

Allium canadense

Allium cernuum

Allium stellatum

Amorpha canescens

Amsonia tabernaemontana v. salicifolia

Andropogon gerardii

Anemone canadensis

Anemone virginiana

Apocynum cannibinum

Arnoglossum atriplicifolium

Arnoglossum plantagineum

Asclepias purpurascens

Asclepias sullivantii

Asclepias verticillata

Astragalus canadensis

Aureolaria grandiflora

Baptisia alba v. macrophylla

Baptisia australis

Baptisia bracteata v. leucophaea

Brickellia eupatoriodes

Camassia scilloides

Carex brevior

Carex davisii

Carex gravida

Carex shortiana

Carex vulpinoidea

Castilleja coccinea
Photo: Mark Berhow

Ceanothus americanus

Chamaecrista fasciculata

Cirsium discolor

Comandra umbellata

Coreopsis palmata

Coreopsis tripteris

Dalea foliosa

Dalea candida

Dalea purpurea

Desmanthus illinoensis

Desmodium canadense

Dichanthelium oligosanthes

Dodecatheon meadia

Echinacea pallida

Echinacea purpurea

Elymus canadensis

Elymus villosus

Elymus virginicus

Eryngium yuccifolium

Eupatorium altissimum

Eupatorium purpureum

Euphorbia corollata

Euthamia graminifolia

Fragaria virginiana

Gaura biennis

Gentiana alba

Gentiana X curtisii (hybrid)

Gentiana puberulenta

Geum triflorum

Helianthus grosseserratus

Helianthus maximiliani

Helianthus mollis

Helianthus occidentalis

Helianthus pauciflorus

Helianthus tuberosus

Heliopsis helianthoides

Heuchera richardsonii

Hypericum punctatum

Hypoxis hirsuta

Iliamna rivularis v. rivularis

Juncus tenuis

Lespedeza capitata

Lilium michiganense

Lithospermum canescens

Lobelia spicata

Lysimachia ciliata

Mirabilis nyctaginea

Oenothera biennis

Monarda fistulosa

Oligoneuron rigidum

Orbexilum onobrychis

Oxalis violacea

Packera paupercula

Panicum virgatum

Parthenium integrifolium

Pedicularis canadensis

Penstemon digitalis

Phlox pilosa

Physalis heterophylla

Physostegia parviflora

Physostegia virginiana

Polemonium reptans

Polygala sanguinea

Polygonatum biflorum v. commutatum

Polytaenia nuttallii

Potentilla arguta

Potentilla simplex

Pycnanthemum tenuifolium

Pycnanthemum verticillatum v. pilosum

Pycnanthemum Virginianum
Photo: Mark Berhow

Ratibida pinnata

Rosa arkansana v. suffulta
Photo: Mark Berhow

Rosa carolina

Rosa setigera

Rosa virginiana

Rudbeckia fulgida

Rudbeckia hirta

Ruellia humilis

Rudbeckia subtomentosa

Salix humilis

Schizachyrium scoparium

Senna marilandica

Silene regia
Photo: Mark Berhow

Silene stellata

Silphium integrifolium

Silphium laciniatum

Silphium perfoliatum

Silphium terebinthinaceum **(large ovate/cordate leaves)**

Sisyrinchium albidum

Sisyrinchium angustifolium

Solidago canadensis

Solidago juncea

Solidago speciosa

Sorghastrum nutans

Sporobolus heterolepis

Spiranthes cernua

Symphyotrichum ericoides

Symphyotrichum laeve

Symphyotrichum novae-angliae

Symphyotrichum oolentangiense

Symphyotrichum pilosum

Teucrium canadense

Thalictrum dasycarpum

Tridens flavus

Tradescantia ohiensis

Tripsacum dactyloides

Vernonia fasciculata

Viola cucullata (likely) or *V. sororia*

Viola pedatifida

Veronicastrum virginicum

Zizia aptera

Zizia aurea

Chapter 3
Dry-Mesic to Dry Soil-Preferring Species

The species outlined in this subsection are found on well-drained soils, such as found on hillsides, sandy gravel, or sands. Important factors are either good drainage or friable soils. For example, friable black prairie loam often supports growth of plants that are normally found in sand or gravelly hillsides.

3.1 **Antennaria howellii* ssp. *neodioica, Antennaria neodioica* (Asteraceae Family), Pussytoes or Smaller Cat's-Foot

Of the 32 North American species of *Antennaria*, *Antennaria howellii* ssp. *neodioica*, pussytoes (height, 0.15–0.3 m), is one of the most common *Antennaria* species found in the tallgrass prairie, as well as in old fields from Maine northwest to Washington, south to North Carolina, and southwest to Colorado, Idaho, and Oregon. In Canada it ranges from Quebec to British Columbia and the Northwest Territory. See illustration on page 141. Another wide-ranging species, **Antennaria neglecta*, is also common. *A. howellii* ssp. *neodioica* is distinguished by the abrupt pointed tip of the basal leaves (see herbarium at Springer website), whereas with *A. neglecta* the obovate basal leaves are not pointed. In addition, *A. howellii* ssp. *neodioica* has leafy stolons throughout whereas *A. neglecta* has leafy stolons only at the ends. Pussytoes is generally found on dry hillsides in full sun or partial shade where there is little competition from other plants. In this regard, it requires a rather specialized location of fairly sterile soil to thrive. If the proper location is selected, its relatively low coefficient of conservatism of 4 ensures success. Flowers appear in early May, and seeds can be collected in late May. Collection should be prompt before strong winds disperse the fluffy seeds. For germination seeds require sowing on top of the soil after 2 months of cool-moist stratification [11].

H.W. Gardner, *Tallgrass Prairie Restoration in the Midwestern and Eastern United States: A Hands-On Guide*, DOI 10.1007/978-1-4419-7427-3_3,

3.2 **Artemisia ludoviciana* ssp. *ludoviciana* (Asteraceae Family), Louisiana Sage or Prairie Sage

Although the flowers of *Artemisia ludoviciana* ssp. *ludoviciana*, Louisiana sage (height, 0.3–1 m), are not remarkable, the plant is conspicuous in prairies for reasons of the whitish-tomentose leaves (illustration on page 141). This subspecies is widespread over the U.S., except West Virginia, Alabama, and Florida. In Canada it is found in Quebec to British Columbia and the Northwest Territory. There are seven other subspecies restricted to either the Southwest or Northwest. Although this plant will survive on mesic soils, it does best on dry-mesic to dry sandy or gravelly soils. In favorable ecosystems, it spreads fairly aggressively by root rhizomes; consequently, its coefficient of conservatism is only 2. Seed-heads can be collected from late October to early November. Seeds plus much extraneous material can be thrashed from the seed-heads after drying. Scraping over quarter-inch hardware cloth removes seed more efficiently but yields far more extraneous material. Greater seed purification can be achieved by sieving through ordinary window screen. Seeds are tiny (about 1–2 mm long) tan cylindrical ellipsoids resembling teardrops. For germination seeds need no treatment, or one can cool-moist stratify for 2 months [11]. The seeds should not be covered with soil [11]. Percentage germination tends to be low.

3.3 **Asclepias tuberosa* (Asclepiadaceae Family), Butterfly Weed

Asclepias tuberosa, butterfly weed (height, 0.3–0.8 m), has a bright yellow-orange to red-orange inflorescence, a rare color in prairies, and thus it is considered very desirable (illustration on page 141). The ssp. *interior* has a wider distribution over most of the U.S., except the upper Northeast and the Northwest, whereas spp. *tuberosa* is restricted to east of the Mississippi River. Blooms appear in early July. It will grow in savannas as well as open prairies. This plant does require fairly restrictive soils, however. This mesic to dry-soil plant requires well-drained friable soils, such as sand/gravel, shale soil, or black prairie silt-loam; sand and silt-loam are best. Black prairie silt-loam, sand, and shale seem contradictory, but it is the loose nature of these soils that seem to be key. Butterfly weed might be started in a nonporous soil, but it will rarely persist. However, certain suppliers offer butterfly weed varieties resistant to clay-like soils. Seed can be collected from the seedpod from mid-August to mid-September when the seed matures to brown; however, the pod can be greenish yellow or greenish tan at this time. One should gently pinch the pod to see if it splits open before harvesting. The pod is generally thinner than other milkweeds and the leaf is also distinctly different, so it is unlikely to mistake it for other *Asclepias* species. For separation of the seed from the silk, see the foregoing procedures recommended for *Asclepias purpurascens* in Chap. 2 (above). The seed, a brown, flat oval (6 mm in length) requires 2 months of cool-moist stratification or no treatment for germination [11]. A cool-moist treatment was found to be successful by Deno [12], who found that germination was 100% in 2–7 days after 3 months cool-moist (40°F) treatment followed by 70°F germination temperature. Light is not a germination factor. Fall planting is also very successful. A coefficient of conservatism in the mid-range indicates a fairly facile propagation provided that the proper soils are available. In friable soil kept free of weed competition, this plant will flower during the first growing season. Caterpillar predation, especially by the Monarch butterfly larva, can be severe.

3.4 **Asclepias viridiflora* (Asclepiadaceae Family), Green Milkweed

Because of its green flower and scarcity in prairies, *Asclepias viridiflora*, green milkweed (height, 0.3–0.9 m), is difficult to locate (illustration on page 141). Green milkweed is found throughout the U.S., except the upper Northeast and the far West. It also occurs in the mid- to western provinces of Canada. Flowering occurs from late June to early July. A high coefficient of conservatism of 9 makes this plant a desirable addition to prairie restoration projects simply for aesthetic reasons. As might be expected, green milkweed is mainly restricted to old, ecologically sound prairies. It is difficult to propagate. This full-sun plant requires dry-mesic to dry hillsides, especially in sandy/gravelly soil. Seedpods (about 6–10 cm long) are smooth, covered with fine down (puberulent), and taper at the upper portion. Flat oval seeds can be collected from the pod and silk when the seeds are a mature brown, from mid-September to mid-October. See *Asclepias purpurascens* in Chap. 2 for seed purification methods. Before harvesting the seeds, pinch the pod to determine if it will readily split open, which is usually a sign of seed ripeness. Experimental germination data [12] showed that cool-moist (40°F) treatment for 3 months followed by 70°F germination temperature was 75% effective within the second day. Light had no effect, and no germination occurred without cold-moist treatment. Accordingly, this author found fall planting to be entirely satisfactory, and it is easy to grow in weeded beds. Monarch butterfly larvae are very common in propagated beds of this species; often the predation is so severe that seed production is nil in some years. Evidently monarchs are attracted to concentrated numbers of green milkweed, while isolated individuals remain largely untouched.

3.5 **Baptisia tinctoria* (Fabaceae Family), Yellow Wild Indigo

Baptisia tinctoria, yellow wild indigo (height, 0.3–0.9 m), is rare in Midwest prairies, but it is offered as a possibility to increase diversity. See illustration on page 141. With a coefficient of conservatism of 10, it is not surprising that yellow wild indigo is on Illinois's endangered species list. In the East it is found from Maine south to Georgia, and in the Midwest it ranges from Ontario and Minnesota south to Iowa, Illinois, and Tennessee. Yellow wild indigo is fairly common in woodland openings in the mountains of the East on soil characterized as either sandy, gravelly, or shale. This author has cultivated the species on Midwestern mesic soil defined as Keomah silt-loam (nearly level, originally forest soil, somewhat poorly drained, brown friable silt-loam), initially giving a spectacular mass of flowers and seedpods. This cultivated specimen did not persist, and permanent restoration failed. Thus, it seems certain that yellow wild indigo does best in mesic to dry soil on sand/gravel/shale and located in either full sun or partial shade. Yellow flowers appear in early July. Because the terminal branches are relatively few-flowered, the black pods (1 cm or less long) are not as notably clustered as the other *Baptisia* species. Seedpods are collected in mid-August to mid-September. The bean-like seeds (2–2.5 mm long) are obtained by crushing the pods. This author has not yet seen weevil infestation within the pods of this species. For germination, seeds should be scarified and given 10 days of cool-moist stratification followed by inoculation with specific nitrifying microorganisms [11]. Fall planting of untreated seeds gives a relatively low percentage of germination in the spring. A low percentage of seeds germinate the second year.

3.6 **Bouteloua curtipendula* v. *curtipendula* (Poaceae Family), Sideoats Grama

Bouteloua curtipendula v. *curtipendula*, sideoats grama (height of seed-head, 0.5–0.8 m) is usually found on dry-mesic or dry hillside prairies or savannas where there is little competition for space with tallgrass or other large plants (illustration on page 141). Although sideoats grama will grow on mesic soil, it is commonly found in profusion on the relatively sterile substrate found on railroad beds. Sideoats grama enjoys wide distribution over most of the U.S. and Canada from Ontario west to British Columbia. It is only absent in a few states of New England, North Carolina, Nevada, and California. When planted in restoration sites, it often flourishes for a short time, but it does not persist if the ecosystem is not favorable. A relatively high coefficient of conservatism of 7 perhaps reflects this requirement for a restrictive environment. Sideoats grama is found in clumps of low fine grass. It flowers in late July to early August. During the initial phase of flowering, this grass has an attractive reddish hue. The oat-like seeds (0.6–1 cm long) can be collected in mid-August to early October, only if the seeds can be easily stripped by hand from the stem. Untoward resistance to stripping means that the seeds are not yet ripe. Similar to most grasses, this species does not require seed treatment for germination [11], but fall planting is an option. Controlled experiments showed 100% germination in 3–6 days at 70°F, and cycling from cool-moist (40°F) for 3 months to 70°F resulted in 97% germination in 2–4 days [12].

3.7 **Callirhoe triangulata* (Malvaceae Family), Clustered Poppy Mallow

Callirhoe triangulata, clustered poppy mallow (height, 0.4–0.7 m), derives its scientific name from the somewhat triangular basal leaves. As seen in the illustration on page 141, it has a bright magenta inflorescence. The dried herbarium specimen shown in the Springer website changed color to a blue violet. This species is found in five upper Midwest states centered in Illinois, and it occurs in the Southeast with the exception of Florida. Clustered poppy mallow has a high coefficient of conservatism of 9, which is a good indication of its difficulty of restoration. Clustered poppy mallow is found in dry-mesic to dry soil in savannas or prairies. Well-drained sandy soil is a virtual requirement, as this plant does not like competition or excess moisture. Poppy mallow will persist only for a few years in the usual non-sand mesic soils. In my experience, too much moisture often triggers root rot. The seeds are borne in typical mallow capsules giving multiple flat, tan "three quarter-moon" bodies that resemble seeds (4 mm). Interestingly, the tan material is actually only a thin-shelled coating that breaks open to reveal a dark-brown "three-quarter moon" seed (3 mm). Seeds are best sown in the late fall. Because the seed coats are rather impervious, germination may not occur the first year. The boiling water method recommended by Prairie Moon Nursery [11] is not advisable unless the water temperature is controlled to be 140°F or less for only 5–10 min [12]. Deno [12] points out that a hole in the seed coat is more effective for germination with seeds having impervious coats.

3.8 **Cirsium hillii* (Asteraceae Family), Hill's Thistle

Cirsium hillii, Hill's thistle (height, 0.2–0.6 m), is considered by at least one authority [6] to be indistinct from the related species, **Cirsium pumilum*; however, other authorities [13, 14] consider it to be separate. Except for their small size, both *C. hillii* and *C. pumilum* are reminiscent of *Cirsium discolor.* See illustration on page 141. According to Fernald [14], *C. hillii* has phyllaries (involucral bracts) with dark glutinous backs with the outer bracts terminated by a slender prickle (1.5–3 mm long), whereas *C. pumilum* mostly lacks the dark glutinous band on the phyllaries and the outer ones are tipped with coarse prickle (4–6 mm long). *C. hillii* is found on Midwestern sites from Indiana and Ontario west to Minnesota, but apparently does not range farther south than Iowa and Illinois. On the other hand, *C. pumilum* is located in more Eastern locales from Maine south to Ohio, Maryland, and North Carolina [14]. The preference of *C. hillii* and *C. pumilum* for dry-mesic to dry hillsides with sandy or gravelly soil also differs considerably from the habitat preferred by *Cirsium discolor*. *Cirsium hillii* is a rare thistle, and it is found almost exclusively on well-established old hill prairies. The assigned coefficient of conservatism of 7 seems a bit low. The seed is borne on a fluffy pappus typical of thistles. Seeds require a 2-month cool-moist stratification for germination [11].

3.9 **Coreopsis lanceolata* (Asteraceae Family), Lance-Leaved Coreopsis

Coreopsis lanceolata, lance-leaved coreopsis (height, 0.25–0.5 m), is an attractive bright-yellow-flowered prairie plant (illustration on page 141). It has been widely cultivated in gardens with the horticultural nomenclature of "garden coreopsis." In the U.S. this species is only absent from the Dakotas, Montana, Idaho, Nevada, Utah, and Arizona. It is also found in British Columbia and Ontario. Although it has a mid-level coefficient of conservatism, it does have an ecosystem preference for dry-mesic to dry sandy or gravelly soil. It does not persist on poorly drained soils. Flowering occurs in mid- to late June. Seed-heads can be clipped when the seed-heads are senescent brown in late June to early August. The brown seeds can be easily recovered from the dried seed-heads, such as scraping over a sieve of quarter-inch hardware cloth. The flat circular seeds are about 4 mm in diameter, which includes the wings on two sides. Recommended seed treatment for germination is either no treatment or 2 months of cold-moist stratification [11]. In my experience, fall planting has proven to be very successful. Precise experiments have shown that the optimum germination was obtained after 6 months dry storage at 70°F, after which 100% of seeds germinate at 70°F within a week [12]. If one minimizes competition from other plants, this plant is easily cultivated in many soils. However, its lifetime is short (about 3–5 years) requiring replenishment by seed. It often decreases as the prairie matures.

3.10 *Delphinium carolinianum* ssp. *virescens* (Ranunculaceae Family), Prairie Larkspur

The flowers of *Delphinium carolinianum* ssp. *virescens*, prairie larkspur (height, 0.4–1.2 m) are white to pale blue, which distinguishes it from the blue flowers of either the mainly woodland species, **Delphinium tricorne* (not shown), or the related more southern species, **Delphinium carolinianum* ssp. *carolinianum* (not shown). Prairie larkspur, ssp. *virescens*, has a requirement for dry-mesic to dry sandy barren prairies (illustration on page 141). It is restricted to the Midwest and eastern Great Plains; that is, in the north, Wisconsin west to Manitoba, the Dakotas, and Colorado, and in the south from Arkansas west to New Mexico. With a coefficient of conservatism of 10, it is difficult to grow. This author's attempt at cultivation in sand was 1 year of flowering before the plants disappeared. Gravelly shale soil was also successful, provided that competing weeds were pulled or an area of barren soil was chosen. The plants may temporarily disappear after an unusually wet season. Prairie larkspur is surprisingly sensitive to hot-dry conditions, but this may be partly the result of early senescence after producing seed. By late July the plants disappear until the next spring. Individual seedpods should be gathered carefully when they senesce to a tan color in late June to mid-July.

Seed drop is very easy, and significant losses can occur during periods of wind or careless handling. Prompt collection without delay is advised. The seeds (about 2 mm long) have flat sides and are notably wrinkled and finely scaled. Seeds should be sown either in the late fall [12], or in the early spring after a 2-month cool-moist stratification [11]. These seeds require cool conditions for optimal germination.

3.11 **Hesperostipa spartea, Stipa spartea* (Poaceae Family), Porcupine or Needle Grass

Hesperostipa spartea, porcupine grass (height, 0.5–1 m), is an early season grass that flowers about late May (illustration on page 142). The preferred habitat is full sun or partial shade in either sandy, gravelly, or silt-loam soil with dry-mesic to dry conditions; however, in the proper friable soil it will thrive under mesic situations. Porcupine grass is located in Canada from Ontario west to British Columbia. In the U.S. its most eastern occurrence is in Pennsylvania. To the west it is found in states north of the Ohio River and west of the Mississippi River to the Rocky Mountains, with the exception of Texas, Arkansas, and Louisiana. The coefficient of conservatism for porcupine grass is 6. Seeds are ripe around mid-June, and they are partially covered with a leafy sheath. The seed can be pulled out of the grain head by its brownish-black fiber that tops the seed or grain. Only those grains that give little resistance to pulling should be harvested. Because the seeds readily drop, harvesting should be on a regular basis. The seed is truly a marvel of botanical engineering. The seed itself (about 2.5 cm long) is reddish brown and somewhat hairy with a very sharp point at the end. This seed is topped by a twisted fibrous tail (about 14 cm long), which includes a sharp bend (about 4 cm long) at the end most distant from the seed. When the seed drops from the plant, the pointed end of the seed penetrates the soil. The hairs on the tip of the seed are recessed backward, ensuring that the seed stays in the soil. The twisted fiber tail responds to humidity by screwing the seed further into the ground. Presumably, the sharp bend at the end serves, by the twisting action, to partly elevate those seeds that initially failed to impale themselves into the soil. It is said that porcupine grass seeds will bore through the noses of grazing cattle. I have experienced the seed penetrating my sock or trousers. After harvest, the mass of seeds will twist their tails into an impossible tangle. Because the fibrous tail responds to moisture, one can untangle them by soaking in water for a short time (about 1–2 h), after which the two fibers become straightened. A successful way of handling the seeds is to simply impale them into the soil soon after harvest (not much depth is required, as the fiber will complete the job). Prairie Moon Nursery [11] recommends that the seeds be given a 2-month cool-moist stratification followed by an early spring planting; however, I have found that pushing the sharp end of the seed into the soil as late as early March will give good results. With large amounts of seed, sowing after harvest is a simpler option.

3.12 **Koeleria macrantha, Koeleria cristata* (Poaceae Family), June Grass

Koeleria macrantha, June grass (height, 0.3–0.6 m), blooms in early June as the namesake implies. See illustration on page 142. June grass is found in most of Canada and the U.S., except parts of the Northeast and most of the Southeast. It is fairly conservative (coefficient of conservatism, 7), and it favors dry-mesic to dry sandy soil in full sun or partial shade. June grass is also found in Eurasia. This grass can be grown on mesic, less well drained soil, but it does not thrive as well. The seed-head can be collected when it is light tan in color around the end of June. The grain (whitish to light tan, about 4 mm by 1 mm) plus chaff can be stripped from the seed-head. For successful restoration, one must remember that June grass prefers cool seasons. Thus, seed should be sowed on top of the soil either late in fall or early in spring; otherwise, no other seed treatment is required [11].

3.13 *Liatris aspera* (Asteraceae Family), Rough Blazing Star

With its purple "button" flowers, *Liatris aspera*, rough blazing star (height, 0.4–1 m), lends an attractive addition to prairies when blooming in late August (illustration on page 142). There are two varieties that share the region of interest. The typical variety, *aspera*, is found in the Midwest and Mid-South. The *intermedia* variety is found in Ontario, New York, the Midwest, Mid-South, and Southeast. Rough blazing star requires well-drained soils in full sun or partial shade. The preferred conditions are often hillsides or sandy/gravelly soil; hence, competition is usually low. It will grow, but not persist, in mesic poorly drained soils. It is debatable whether this lack of persistence is caused by competition with other plants, a short life in wet soil, or predation by rodents, the latter of which prefer to reside in lush vegetation. During winter, rodents will decimate *Liatris aspera* by eating the tuberous root or corm. The coefficient of conservatism is rated at 7. The seed ripens on individual buttons from the top down at different times from mid-September to late October. Seeds are ready to collect when the individual buttons become fluffy with tan pappus. Because the seeds are eaten by birds or "gone with the wind" when the pappus fluffs out, many visits for more than a month are necessary to efficiently collect most of the seeds. The seed is brown (4–6 mm long, 1 mm wide where it attaches to the pappus, then narrowing to a point), and the fluffy pappus is 6–8 mm long. Seeds require a 2-month cool-moist stratification for germination [11]. Sowing seeds in the fall is also a successful method.

3.14 **Liatris cylindracea* (Asteraceae Family), Cylindrical Blazing Star

Liatris cylindracea, cylindrical blazing star (height, 0.2–0.5 m), is superficially similar in appearance to *L. aspera*, and especially to *L. squarrosa* (illustration on page 142). Both *L. cylindracea* and *L. aspera* have appressed floral bracts, whereas

L. squarrosa has lanceolate bracts that curl downward as the flower senesces. The floral bracts of both *L. aspera* and *L. cylindracea* are ovate or oblong; only *L. cylindracea* bracts have pointed tips. Both *L. cylindracea* and *L. squarrosa* have cylindrical or turbinate flower heads, but *L. aspera* has more rounded floral heads. Cylindrical blazing star is also generally lower in height than *L. aspera. L. cylindracea* requires dry-mesic to dry soils in the full sun, such as found on sterile hillsides or sandy soil. This plant does not persist on mesic or poorly drained soil unless intensively cultivated to eliminate competition. It is mainly located in the Midwest, but it has been reported in New York, Ontario, Delaware, and Alabama. A coefficient of conservatism of 8 is indicative of a difficult restoration. Seeds can be collected when the tan pappus fluffs out from late September to early October. The seeds and pappus resemble those of *L. aspera*, except they are generally larger (seeds, 6–7 mm long; pappus, 7–10 mm). For germination, seeds require 2 months of cool-moist stratification [11]. Alternatively, the seeds can be fall planted.

3.15 **Liatris squarrosa* v. *squarrosa* (Asteraceae Family), Scaly Blazing Star

Liatris squarrosa v. *squarrosa*, scaly blazing star (height, 0.15–0.6 m), represents one of many *Liatris* species with purple flowering "buttons." See illustration on page 142. *L. squarrosa* resembles *L. cylindracea*, and to some extent *L. aspera* (see *Liatris cylindracea* section, above). Among the above three, *L. squarrosa* is unique with its lanceolate bracts, which turn downward as the flower senesces. *L. squarrosa* requires dry-mesic to dry soils, but in my hands drought is hard on this species. The key to its survival may be good soil drainage. Scaly blazing star grows best in full sun or partial shade. There are five varieties of scaly blazing star, of which only v. *squarrosa* inhabits both the East (Maryland to the south) and the Midwest (Michigan and Ohio southwest to Texas). A coefficient of conservatism of 7 indicates little tolerance for disturbance. Seed-heads can be collected when brownish-tan fluff appears from early September to early October. At this time the seed, fluffy pappus, and bracts readily separate from the clipped seed-head. A 2-month cool-moist stratification is needed for germination [11]. Fall planting gives a fair degree of success.

3.16 **Lithospermum caroliniense* v. *croceum* (Boraginaceae Family), Hairy Puccoon

Lithospermum caroliniense v. *croceum*, hairy puccoon (height, 0.2–0.75 m), is similar to *L. canescens*, except it tends to be a larger plant, more leafy, with longer hairs on the leaves (not hoary). See illustration on page 142. According to *Gray's Manual* [14], there are sufficient differences in the more southerly species, *L. caroliniense*, compared to the more northerly *L. croceum* to consider them as separate species; however, Barkley et al. [15] have listed the two as the same. Currently, the USDA website has it listed as a variety, *L. caroliniense* v. *croceum*. This variety has a northern limit from New York west to Ontario, Minnesota, South Dakota, and Colorado. Its southern limit is Pennsylvania, West Virginia, Tennessee, Arkansas, and Oklahoma. Hairy puccoon is likely to be found on dry-mesic to dry sand barrens. It is a savanna or prairie species that flowers in late May. The coefficient of conservatism is 7. The seeds, ovoid white nutlets about 4 mm long, are ripe in August. Specific seed treatment for germination is not known, but suggestions are given for the related *L. canescens* above in Chap. 2.

3.17 ******Lupinus perennis* ssp. *perennis* (Fabaceae Family), Wild Lupine

Lupinus perennis ssp. *perennis*, wild lupine (height, 0.2–0.6 m) affords very attractive blue blossoms in mid- to late May, and other forms give pink or white flowers (illustration on page 142). Its favored habitat is either full sun or partial shade on well-drained dry-mesic to dry soils, preferably with friable, sandy/gravelly character. There are three subspecies; the typical ssp. *perrenis* comprises two varieties that share similar areas. The subspecies *perennis* ranges in the north from Newfoundland and Maine west to Ontario and Minnesota. From the northern limit it ranges south and east to Iowa, Illinois, Kentucky, and along the East Coast from Virginia to Georgia. One other subspecies is found in the Southeast and Mid-South. For several years, this author has grown lupine in central Illinois on dry-mesic hillsides with less friable "forest" soils, and a few plants persisted for several years. Lupine grows well in shale soils in southeastern Pennsylvania, but lupine does best in sand or soil amended with sand or gravel. Poor soils undoubtedly are required to reduce competition from other plants, but lupine does not fare well in periods of hot drought. Rated at 8, its coefficient of conservatism is relatively high. In mid-June four to six whitish to tan seeds (4 mm) are borne in pods (8 mm by 3–4 cm long) covered with down-like hairs (pubescent). For germination seeds should be scarified followed by 3 days of cold-moist stratification and then inoculated with specific nitrifying bacteria [11]. It seems important to place a hole in the seed coat to trigger germination. If fall planted, seeds will germinate at low percentage without any treatment. This is an early spring plant; thus, seeds should be placed in the soil very early.

3.18 **Mimosa microphylla, Schrankia uncinata/Schrankia nuttallii* (Fabaceae Family), Sensitive Brier or Cat-Claw

The most notable characteristics of *Mimosa microphylla*, sensitive brier or cat-claw (length, 0.6–1.2 m, decumbent or prostrate), are the folding of the leaf to touch, as well as the prickles on the stem and seedpod (illustration on page 143). Sensitive brier prefers areas of low competition in full sun with sandy or gravelly dry to dry-mesic soil. It thrives best in sand. Because of its restrictive habitat, it is not too common. Its coefficient of conservatism is 7. Its range includes Virginia and Kentucky south to the states in the Gulf. It is also found in Illinois. Magenta globose flowers appear in early July and may continue blooming until the first heavy frost. Seeds are contained in prickle-covered pods, which can be collected from late August until mid-October. The pods should be picked only when brown; the seeds from greenish-brown pods are not yet ripe. Loss of seeds occurred when the pods were permitted to remain too long after senescence. Upon drying in paper bags the pods will open to release the seeds. Recommended seed treatment is scarification of the seed followed by 10 days of cool-moist stratification [11]. I have found that fall planting without seed treatment was successful. A few seeds may germinate the second growing season.

3.19 **Monarda punctata* ssp. *punctata* (Lamiaceae Family), Spotted Bee Balm

Monarda punctata ssp. *punctata*, spotted bee balm (height, 0.2–0.8 m), has been variously described as annual, biennial, or perennial. If perennial, it is probably short lived. See illustration on page 142. This plant's ecological requirement is dry-mesic to dry sandy soil in full sun or partial shade, but it is surprisingly intolerant of midsummer hot drought. If the conditions are right, it apparently is easy to restore (coefficient of conservatism, of 5). This author has successfully row-cropped spotted bee balm in mesic poorly drained soil with a good seed yield initially, but the plant rapidly disappeared and did not respond to restoration on such soils without cultivation. Also, the plant responded to cultivation on gravelly shale soil, but disappeared by the third year. It is likely that sandy soil is optimal. For example, it grows wild on the sand hills east of the Illinois River. The subspecies *punctata* comprises seven varieties. The *punctata* subspecies has a northern limit from Vermont west to Quebec, Ontario, Minnesota, and Kansas. The southern extent stretches from Florida west to New Mexico. It is also found in California. One other subspecies in located only in Texas. Seedheads can be clipped when senescent brown from early September to mid-October. The reddish-brown or brown seeds are oblong (1.2–1.5 mm long). Seed collection follows the procedures outlined for *M. fistulosa* described above. Seeds require no treatment for germination but should be sowed on top of the soil [11]. However, Deno [12] determined more precise conditions. According to him, seeds dry-stored for 6 months germinated 51% at 70°F in the light in the second week, but in the dark at 70°F germination was only 4%. A 3-month cool-moist treatment followed by 70°F in the light resulted in 39% germination within 2 weeks, but a similar dark treatment gave only 8% germination. Seeds have a half-life of about 1 year.

3.20 **Onosmodium bejariense* v. *occidentale*, *Onosmodium molle* v. *occidentale* (Boraginaceae Family), Western False Gromwell

Onosmodium bejariense v. *occidentale*, western false gromwell (height, 0.4–1 m), is one of three similar species or varieties. An early reference [13] and a more recent one [14] categorized the three as separate species; another [15] lists *O. molle* as a species with two additional varieties, v. *hispidissimum* and v. *occidentale*. Yet another reference [6] names these varieties as *O. molle*, *O. molle* v. *occidentale*, and *O. hispidissimum*. Currently, the USDA website lists them as varieties: *O. bejariense* v. *bejariense*, *O. bejariense* v. *occidentale*, and *O. bejariense* v. *hispidissimum*. The variety shown here is best described as its current name *Onosmodium bejariense* v. *occidentale* (http://plants.usda.gov/java/nameSearch) (illustration on page 143). It is a rather rough, hairy plant that is usually found in sandy/gravelly soils in mesic to dry conditions in full sun or partial shade. Even though its preference is sand/gravel, it does hold its own in less friable soils on hillsides or drier situations. It is located in the U.S. in the Midwest and Plains to the Rocky Mountains, as well as southern-mid Canada. It is also found in the Southeast, except for the Carolinas and Florida. Its coefficient of conservatism is a fairly high 8. Western false gromwell blooms with rather inconspicuous greenish flowers in early July, and sets individual whitish nutlets (about 4 mm long) that can be harvested from early September to mid-October. The nutlet seeds distinguish the varieties as follows: *O. bejariense* v. *bejariense*, nutlet distinctly pitted (2 mm long) and slightly constricted at the base; **O. bejariense* v. *hispidissimum*, nutlet distinctly constricted at the base, not pitted or little pitted (3 mm long); **O. bejariense* v. *occidentale*, nutlet scarcely or not pitted, not constricted at the base (4 mm long). Because of the hard seed coat, various seed treatments to stimulate germination have been suggested. Fall planting is one option used by this author, with most of the seeds germinating a full year from the following growing season. Reportedly, seeds can take 5–8 years to germinate [12]. Another option is seed scarification followed by a 2-month cool-moist stratification. A third option is soaking for 1 day, followed by 1–2 months at room temperature sealed in a plastic bag, then a 2-month cool-moist stratification; this latter condition would somewhat mimic the conditions for the observed germination in the second year.

3.21 **Opuntia humifusa*, *Opuntia compressa* (Cactaceae Family), Western Prickly Pear Cactus

Opuntia humifusa, western prickly pear cactus (height, 0.1–0.15 m, prostrate), is found on dry sandy soil in full sun or partial shade distributed from Massachusetts west to Minnesota, South Dakota, and Colorado and south of this northern area to the border. It is also found in Ontario and Montana. Its spectacular yellow flowers in mid-June to early July makes this plant an attractive addition to your landscape, even if sand has to be imported into your project (illustration on page 143). It does not respond well to prairie fires. Fuel is not usually abundant in its native habitat. It will slowly recover from a fire. Its coefficient of conservatism is 5. Club-shaped fruit (2–5 cm long) change from green to red or purple. When the fruit is red or purple in early September, it can be broken open to collect spherical seeds somewhat flattened on two sides. The tan seeds have a narrow rim and measure 4.5–5 mm in diameter. The seeds need no treatment for germination [11]; however, it is my experience that germination can be very unreliable. As a result of qualitative experiments, Deno [12] found some immediate germination at 70°F. Breaking off a cactus frond/leaf at the joint and inserting the broken end in sand is an alternative and facile method of propagating this cactus. It will eventually spread into a large matted colony.

3.22 *Penstemon grandiflorus* (Scrophulariaceae Family), Large-Flowered Beard-Tongue

Penstemon grandiflorus, large-flowered beard-tongue (height, 0.6–1.2 m), is restricted to dry-mesic to dry prairies and savannas (illustration on page 143). This plant is tolerant of drought, and I have found that it will thrive only on sand. Small plants will set seed growing on gravelly shale, but their development is not optimum and eventually the plants disappear. Probably because of its restrictive requirements for well-drained sand, it is endangered in Illinois and has been given a coefficient of conservatism of 8. Its northern range is from Ohio and Michigan west to Montana. In the east it is found in Massachusetts and Connecticut. In the Midwest it ranges north of the Ohio River and west of the Mississippi River to the Rocky Mountains, except for its absence in Arkansas and Louisiana. This beard-tongue is especially attractive, not only for its large lavender flowers, but also for its thick-fleshy leaves that are glabrous (smooth) and glaucous (whitish bloom). The rather large seed capsules (1.6–2.5 cm), being less abundant in number than those of *P. digitalis*, can be picked individually in mid-August as they turn brown. The hard-shelled capsule can be broken to obtain angular brown seeds (2.5–4 mm long). For germination the seeds require an early spring planting after a 1-month cool-moist stratification [11]. This author had success by using a fall sowing on sand followed by barely covering the seeds with sand. Precise measurements by Deno [12] showed that seeds need dry storage for 6 months at either 70 or 40°F, followed by germination at 40°F (100% germination at 6% per day with an induction time of 43 days). Fresh seeds did not germinate at 70°F, and germinated poorly with a 3-month cool-moist treatment followed by 70°F (3%). These latter data argue against fall planting. However, after seed collection in August the seeds are usually stored for 4 months before a late fall seeding; thus, one might rationalize that a fall sowing on sand might partly satisfy a 6-month warm or cool dry storage, after which the seeds would lie dormant for an additional few months before cool germination in March or April.

3.23 *Phlox bifida (Polemoniaceae* Family), Sand or Tenpoint Phlox

Phlox bifida, sand or tenpoint phlox (height, 0.1–0.2 m), is a dry sand dweller, and in this habitat it grows in full sun or partial shade (illustration on page 143). There are two subspecies, ssp. *bifida* and ssp. *stellaria*, with somewhat overlapping occurrence. The widest ranging ssp. *bifida* occurs in ten Midwestern states centered in Illinois and Missouri. With a coefficient of conservatism of 7 and its restrictive habitat, it is not common. Attractive blossoms are found in late May. The seeds are contained in an oblong spherical capsule (2–3 mm long), and seed shatter (seed loss by natural scatter) is a problem for seed collection. For *Phlox* species it is often recommended that nylon stockings or other capture device should be placed over the capsules to catch the tiny dark seeds when they ripen and shatter. For germination, the seeds need a 2-month cool-moist stratification. Dry storage of the seeds is fatal [12].

3.24 **Polygala senega* (Polygalaceae Family), Seneca Snakeroot

In contrast to *P. sanguinea* described above in Chap. 2, *Polygala senega*, Seneca snakeroot (height, 0.1–0.5 m), is a perennial plant, which is often found growing in colonies (illustration on page 143). The name derives from its use by the Seneca Indians. Seneca snakeroot is usually found on dry-mesic to dry calcareous gravelly soils, often on hillsides or railroad right-of-ways. The plant prefers full sun or partial shade. Seneca snakeroot is found in all southern provinces of Canada. In the U.S. it is located from the East to the Great Plains, with the exception of the states in the Gulf. Its coefficient of conservatism is rated at 7. White flowering spikes, which can be observed in late May, develop into plump round capsules (2.5–4.2 mm long). Seed-bearing capsules drop off the lower portion of the flowering spike. Because this plant can be inconspicuous after blooming, it is advisable to flag the plant or colony for later seed collection. The black seeds are sparsely hairy (2.5–3.5 mm long). Prairie Moon Nursery recommends a 2-month cool-moist stratification for seed germination [11]. Deno [12] tested various germination conditions, but he cautioned that the seeds germinated at 40°F failed to develop after shifting to 70°F. The best conditions he reported were alternating 3-month moist treatments at 70°F in the light, 40°F, and then 70°F in the dark, which gave 41% germination in 2–6 days.

3.25 *Psoralidium tenuiflorum, Psoralea tenuiflora* (Fabaceae Family), Scurf-pea

Psoralidium tenuiflorum, scurf-pea (height, 0.4–1 m), is a bushy legume with a high coefficient of conservatism of 8 (illustration on page 143). In the U.S. scurf-pea is found mainly west of the Mississippi River, except for its absence in Arkansas, Louisiana, and the five most western states. East of the Mississippi River it occurs in Mississippi, Kentucky, Indiana, Illinois, and Wisconsin. Its preferred habitat is dry-mesic to dry soils in full sun or partial shade, often on hillsides with sandy or gravelly soil. It will grow on less porous soils if drainage is adequate. Animals, presumably deer, graze on the plant, but it recovers. Scurf-pea blooms in late June, and the oval brown-dotted pods (5–8 mm long) ripen from late July to late August. Because the pods ripen at various times over 1-month period, many visits are required to harvest the ripe pods before they drop. One seed (olive-brown, smooth and oval, about 4 mm long) is found when the pod is broken open. For germination, the recommended seed treatment is scarification, 10 days of cool-moist stratification, and inoculation with specific nitrifying microorganisms [11]. Less than desirable germination was obtained by a fall planting.

3.26 **Solidago nemoralis* v. *nemoralis* (Asteraceae Family), Old-Field Goldenrod or Dyers Weed

Solidago nemoralis v. *nemoralis*, old-field goldenrod (height, 0.15–1 m), prefers to grow in dry-mesic to dry soils in full sun or partial shade. See illustration on page 143. The typical variety, *nemoralis*, has a northern limit from Nova Scotia west to Manitoba and North Dakota. In the U.S. it is found from the East Coast to the eastern Great Plains, except for South Dakota and Nebraska. Its coefficient of conservatism is a low 3, indicating a tolerance of disturbance. In one old field it was dominant, until it was easily replaced by *Sorghastrum nutans*. Blooms appear in late August. The seed-heads can be clipped when the fluffy pappus becomes evident in late October. The small whitish to brownish seed plus fluffy pappus can be thrashed from the seed-head when dry. For germination the seeds should be sowed on top of the soil in the late fall, or cool-moist stratified for 2-months [11].

3.27 **Spiranthes tuberosa, Spiranthes beckii* (Orchidaceae Family), Little Ladies' Tresses

A comparatively smaller *Spiranthes* species, *Spiranthes tuberosa*, little ladies' tresses (height, 0.13–0.23 m), blooms in late August (illustration on page 144). The basal ovate leaves are senescent (absent) by the time of flowering. Before flowering it is virtually impossible to locate the plant in the field. The flower is much smaller and delicate compared to *Spiranthes cernua* described above. When the flower senesces, tiny seeds can be shaken loose from the brown seed-head in late September. *Spiranthes tuberosa* is found in dry-mesic to dry soil in full sun or partial shade. This member of the orchid family has a high coefficient of conservatism of 9. Surprisingly, it volunteered in an old field on my property in south-central Pennsylvania. Although this is mainly a southern species, it ranges north into Massachusetts, New York, and west to Michigan, southern Illinois, Missouri, and Kansas. The very fine seeds should be scattered on top of the soil in the fall.

3.28 **Sporobolus compositus* v. *compositus*, *Sporobolus asper* (Poaceae Family) Rough Dropseed

Sporobolus compositus v. *compositus*, rough dropseed (height, 0.3–1 m), has thin curly leaves, and it prefers full sun with dry-mesic to dry conditions, generally on sandy, gravelly, or friable silt-loam soil. See illustration on page 144. If well-drained friable soil is present, it will grow in mesic soil. Rough dropseed is found in Canadian provinces from Quebec to Manitoba, as well as British Columbia. In the U.S. it is found in all states, except California, Nevada, and the three most southern states on the East Coast. Two other varieties are found localized in the Mid-South. The coefficient of conservatism is a low 3, but of course this applies to a somewhat restrictive soil requirement. A leafy sheath that curls around the seed-head often conceals the seeds. Seed-heads can be clipped from late September to mid-October. Subjecting the seed-heads either to shuffling action underfoot or scraping on quarter-inch hardware cloth is sufficient to loosen the seeds. The most notable feature of the "seed" is a light tan appearance of a chaffy covering that hides the smaller oval brown grain, measuring about 1.3–2.0 mm long. In a batch of harvested "seed," the chaff mainly migrates to the top with heavier detached small grain sinking to the bottom. Separating the straw from the seed plus chaff requires sieving through quarter-inch hardware cloth followed by sieving through ordinary window screen. For germination, no treatment is required [11].

3.29 **Symphyotrichum oblongifolium*, *Aster oblongifolius* (Asteraceae Family), Aromatic Aster

At the first impression *Symphyotrichum oblongifolium*, aromatic aster (height, 0.2–0.8 m), has the appearance of a smaller version of *S. novae-angliae*, but it is quite different in both habitat and difficulty in restoration (coefficient of conservatism, 7). See illustration on page 144. Its range extends in the north from New York west to Montana. Aromatic aster has been documented in all states south and east of Montana; however, it is absent in New England, New Jersey, Delaware, Louisiana, and South Carolina, Georgia, and Florida. This bluish-purple aster prefers full sun with good drainage, such as sandy or gravelly soil on hillsides. Consequently, dry-mesic to dry conditions are recommended. This author has grown aromatic aster on relatively poorly drained mesic soil, but these plants have not either grown substantially or spread. This aster blooms in late September, but the seed-heads are among the last of the prairie plants to be collected in early November to early December. The seed plus pappus are recovered from the seed-head by procedures described in Chap. 2 for *S. oolentangiense*. The seed plus pappus (about 5 mm long) reportedly requires no treatment for germination [11]. According to Deno [12], dry storage at 70°F for 6 months is required to achieve germination of 100% at 70°F in the second week and at 40°F in 3–5 weeks. Fresh seed required cycles of 40 and 70°F (3 months each) to complete germination [12].

3.30 *Symphyotrichum sericeum, Aster sericeus* (Asteraceae Family), Silky Aster

Symphyotrichum sericeum, silky aster (height, 0.3–0.6 m), has a rather restrictive habitat of dry-mesic to dry hillsides, preferably of sandy or gravelly character with little competition from other plants (illustration on page 144). During wet periods, the plant appears to suffer from fungal infections. It is identified by its silvery-silky leaves, which gain their appearance by a covering of fine hairs. Normally, this aster has purple to blue flowers, but the white form (forma *albiligulatus*) is somewhat less common (see color herbarium on Springer internet site). Silky aster is principally a plant of the Midwest, ranging from Ontario west to Manitoba in the north and ranging south through the Dakotas, Nebraska, and Kansas to New Mexico. Eastward its range extends to the Mississippi River, except Louisiana. It is also found in all states north of the Ohio River to its U.S. eastern limit in Ohio. It has a high coefficient of conservatism of 9, indicating its difficulty of restoration. Flowers appear in late fall requiring seed collection in late October to late November. Seeds require 2 months of cool-moist stratification for germination [11]. In either shale soil or sand this aster was easy to grow from a fall planting in weeded beds.

3.31 **Tephrosia virginiana* (Fabaceae Family), Goat's Rue

The legume *Tephrosia virginiana*, goat's rue (height, 0.3–0.6 m), displays an attractive three-toned flower with yellow, white, and purple/rose around the end of June (illustration on page 144). Unfortunately, this plant is restricted to dry or dry-mesic sands or gravel. Although it will easily grow in full sunlight, it is often found in open woods and savannas. Its northern range is from New Hampshire and New York west to Ontario, Minnesota, and Nebraska. From the north it is located in all states south and east. A coefficient of conservatism of 7 also restricts the plant to fairly undisturbed habitats. When the seedpods are a senescent brown, they can be harvested in early to late September; however, the pods shatter soon after ripening. The seeds may be ballistic as indicated by the curled shattered pods, much as observed with *Chamaecrista fasciculata*. For germination the seeds should be scarified, and stratified under cool-moist conditions for 10 days [11]. According to Deno [12], dry storage caused the seed coats to become more impervious to water. Thus, fresh seeds germinated 100% at 70°F (in 1–10 weeks), but seeds dry stored for 6 months at 70°F germinated only 16% over 3 weeks. Dry stored seeds can be made to germinate 100% if a hole is ground through the seed coat. Seeds dry stored for 1 year germinated similar to 6 months dry storage, but 3 years of storage caused seed death.

3.32 *Tradescantia bracteata* (Commelinaceae Family), Prairie Spiderwort

Tradescantia bracteata, prairie spiderwort (height, 0.1–0.4 m), is listed as an endangered plant in Illinois; however, its coefficient of conservatism is a fairly low 7 for an endangered species. See illustration on page 144. Its northern limit stretches from Michigan west to Montana. Its southern limit is Indiana west to Illinois, Missouri, Oklahoma, and north to Nebraska and Wyoming. Its eastern presence in Massachusetts and Vermont may be the result of importation. Even though it is a fairly small plant, its bright green leaves and brilliant rose-colored flowers attract attention. Blue- and blue/pink-flowered forms are also found (see color herbarium on Springer website). The defining characteristic of *T. bracteata* is the glandular-viscid pubescence of the sepals, which become notable when collecting seed. Flowers appear in mid-May. In the absence of competition this plant will bloom into the late summer. Its habitat is sandy or gravelly dry-mesic to dry prairies. It will hold its own in mesic nonsandy soil, but perhaps competition with other larger plants in these situations may be a factor. Prairie spiderwort spreads slowly by rhizomes in a nonaggressive manner. Two to six seeds per fruit can be harvested by picking the capsules in late June to early July. Because the seeds shatter readily from brown capsules, it is advisable to pick the seed capsules when yellowish green. The oblong seed is 2–4 mm long. Seeds should be planted in late fall. Alternatively, it is suggested that seeds should be given a 4-month cool-moist stratification followed by planting in early spring [11]. Germination experiments [12] showed that a 3-month cool-moist (40°F) period followed by 70°F in the dark gave 50% success within 10 days. Once established, this species can be spread to new areas by transplanting part of the rhizomatous colony. As this is an Illinois listed species, digging should not be done in the wild. The plant or seed can be obtained from reputable nurseries.

A similar species comprising three varieties, *Tradescantia occidentalis* (not shown), with smaller petals and sepals is found west of western Wisconsin and the Mississippi River to the Rocky Mountains. Apparently, it has been imported into New York and New Jersey.

3.33 **Verbena stricta* (Verbenaceae Family), Hoary Vervain

Verbena stricta, hoary vervain (height, 0.3–1.2 m), is a plant with a fairly high tolerance of disturbance (coefficient of conservatism, 2). Hoary vervain is a rather rough plant, but it can be attractive in full bloom (illustration on page 144). It is commonly found on railroad roadbeds, where it gains a competitive edge. Hoary vervain prefers full sun or partial shade in dry to dry-mesic soil. This species occurs in Canadian Quebec and Ontario. In the U.S. it is found in nearly every state, with the exceptions of upper New England, Oregon, California, Louisiana, Florida, South Carolina, Virginia, and Maryland. Hoary vervain does have attractive flowers appearing in mid-August. Purple (common), rose, or white floral forms are known. Differing from *Verbena hastata*, *Verbena stricta* is an authentic perennial according to authorities as well as my observation. Seed-heads can be clipped when senescent brown in late September. Seeds and extraneous material can be recovered by scraping the seed-heads over quarter-inch hardware cloth used as a rough sieve. Grayish-brown seeds are 2–3 mm long. Germination requires 2 months of cool-moist stratification followed by sowing on top of the soil [11]. Sowing in the fall is an easy alternative.

3.34 **Viola pedata* (Violaceae Family), Birdfoot Violet

Viola pedata, birdfoot violet (height, 6–15 cm), can be observed either as the all lilac-violet-flowered form or as the form shown on page 144 with very dark violet upper petals and lilac-violet lower petals. In contrast to the similar-appearing *V. pedatifida*, *V. pedata* does not produce any cleistogamous flowers (small self-pollinating flowers that remain closed). Also, another distinguishing feature is the conspicuous orange-tipped stamens in the center of the *V. pedata* flower. This species prefers sand and sandy-gravelly soils (dry-mesic to dry) either in full sun or partial shade. Grown on nonsandy soil this species does not persist for long. In the north it is located from Maine and Ontario west to Minnesota and Nebraska. Its southern range is south and east of Nebraska, except Florida. It is fairly conservative with a coefficient of conservatism of 7, possibly because of its restrictive soil requirements. Birdfoot violet blooms in late May, and the oblong capsules can be clipped when they first turn yellowish-brown in late June. When the capsules dry, they will split open and release the seeds (egg-shaped and reddish, tan or copper-colored, 1.4–1.7 mm long). Recommended seed treatment for germination is 2 months of cool-moist stratification or fall sowing on top of the soil [11].

Mesic/adaptive species tolerating both dry-mesic and dry soils follow: *Allium stellatum, Amorpha canescens, Amsonia tabernaemontana, Andropogon gerardii, Asclepias verticillata, Baptisia alba macrophylla, Baptisia bracteata leucophaea, Brickellia eupatorioides, Carex brevior, Carex gravida, Coreopsis palmata, Dalea candida, Dalea purpurea, Desmodium illinoense, Dichanthelium oligosanthes, Echinacea pallida, Elymus canadensis, Euphorbia corollata, Fragaria virginiana, Gentiana puberulenta, Geum triflorum, Helianthus occidentalis, Heuchera richardsonii, Juncus tenuis, Lespedeza capitata, Lithospermum canescens, Oenothera biennis, Oligoneuron rigidum, Panicum virgatum, Pedicularis canadensis, Phlox pilosa, Potentilla arguta, Potentilla simplex, Rosa arkansana, Rosa carolina, Rudbeckia hirta, Ruellia humilis, Salix humilis, Schizachyrium scoparium, Sisyrinchium albidum, Solidago juncea, Solidago speciosa, Sorghastrum nutans, Sporobolus heterolepis, Symphyotrichum ericoides, Symphyotrichum oolentangiense, Symphyotrichum pilosum, Tradescantia ohiensis,* and *Tridens flavus.*

Those mesic/adaptive species tolerating dry-mesic soils are *Allium canadense, Allium cernuum, Anemone virginiana, Arnoglossum atriplicifolium, Asclepias purpurascens, Astragalus canadensis, Aureolaria grandiflora, Carex vulpinoidea, Castelleja coccinea, Chamaecrista fasciculata, Cirsium discolor, Ceanothus americanus, Comandra umbellata, Coreopsis tripteris, Dalea foliosa, Desmanthus illinoensis, Desmodium canadense, Dodecatheon meadia, Elymus villosus, Eryngium yuccifolium, Eupatorium altissimum, Eupatorium purpureum, Euthamia graminifolia, Gaura biennis, Gentiana alba, Helianthus pauciflorus, Helianthus maximiliani, Helianthus mollis, Heliopsis helianthoides, Hypericum punctatum, Hypoxis hirsuta, Lobelia spicata, Mirabilis nyctaginea, Monarda fistulosa, Orbexiium onobrychis, Oxalis violacea, Packera paupercula, Parthenium integrifolium, Penstemon digitalis, Physalis heterophylla, Polemonium reptans, Polygala sanguinea, Polygonatum biflorum, Polytaenia nuttalii, Pycnanthemum tenuifolium, Pycnanthemum verticillatum pilosum, Ratibida pinnata, Rosa setigera, Rosa virginiana, Rudbeckia fulgida, Senna hebecarpa, Senna marilandica, Silene regia, Silene stellata, Silphium integrifolium, Silphium laciniatum, Silphium terebinthinaceum, Sisyrinchium angustifolium, Solidago canadensis, Spiranthes cernua, Symphyotrichum laeve, Symphyotrichum novae-angliae, Viola affinus/sororia, Viola pedatifida*, and *Zizia aptera.*

Dry-Mesic to Dry Preferring Species

Antennaria howellii ssp. neodioica

Artemisia ludoviciana

Asclepias viridiflora

Bouteloua curtipendula

Asclepias tuberosa

Baptisia tinctoria

Callirhoe triangulata

Cirsium hillii

Coreopsis lanceolata

Delphinium carolinianum ssp. virescens

Hesperostipa spartea

Koeleria macrantha

Liatris aspera

Liatris cylindracea

Liatris squarrosa

Lithospermum caroliniense v. croceum

Lupinus perennis

Monarda punctata ssp. punctata

Mimosa microphylla

Onosmodium bejariense v. occidentale

Penstemon grandiflorus

Opuntia humifusa

Phlox bifida

Polygala senega

Psoralidium tenuiflorum

Solidago nemoralis

Spiranthes tuberosa

Sporobolus compositus

Symphyotrichum oblongifolium

Symphyotrichem sericum
Photo: Douglas G. K. Franks

Tephrosia virginiana

Verbena stricta
Photo: Mark Berhow

Tradescantia bracteata

Viola pedata

Chapter 4
Wet-Mesic to Wet Soil-Preferring Species

4.1 ***Angelica atropurpurea* (Apiaceae Family), Angelica or Alexanders**

Angelica atropurpurea, angelica (height, 1–3 m), is generally found in bottomland savannas; however, it also can be found in wet-mesic prairies (see illustration on page 158). Because of its large leaves, stout purple stem, and height, it gives a striking and attractive aspect. It is distributed from Maine west to Minnesota, and in the south from North Carolina west to Iowa. In the Canadian provinces it is found from Newfoundland west to Ontario. The seeds should be collected when matured to a tan-brown from late June to mid-July. By mid- to late July the plant becomes senescent. To achieve optimal germination, Prairie Moon Nursery recommends alternating 2-month periods of cool-moist, warm-moist, and cool-moist [11]. However, fall planting in weeded beds with occasional irrigation has proved to be somewhat successful. Although angelica has been assigned a Coefficient of Conservatism of only 6, this seems to be a bit low as the plant is not too common. Angelica does thrive on creek bottoms and other locations considered as "wasteland," which may account for its relatively low rating. Authorities consider this species as a "perennial," but in my hands the plant dies after going to seed (fruit). Angelica usually requires a few years of growth before reaching fruiting stage.

4.2 ***Asclepias incarnata* ssp. *incarnata*, Swamp Milkweed**

Asclepias incarnata ssp. *incarnata*, swamp milkweed (height, 0.3–1.5 m), is found in wet to mesic soil in full sun, but prefers the more moist habitats. In late July its pink to rose-purple flowers attract attention. See illustration on page 158. The ssp. *incarnata* is found all over the U.S., except South Carolina, Mississippi, and the West Coast. In Canada it is found from Quebec west to Manitoba. One other subspecies, ssp. *pulchra*, is distributed along the East Coast and parts of the Midwest. As indicated by its Coefficient of Conservatism of 4, swamp milkweed is easy to restore in wet areas, and it will volunteer if a seed source is nearby. The follicles (pods) contain-

H.W. Gardner, *Tallgrass Prairie Restoration in the Midwestern and Eastern United States: A Hands-On Guide*, DOI 10.1007/978-1-4419-7427-3_4,

ing the seeds are ready for collection in mid-September to early October. The seeds can be purified by the procedure described for *Asclepias purpurascens* in Chap. 2. Deno [12] has found that seed germination absolutely required light and a temperature cycle: 100% germination occurred within 1–3 weeks after 3 months of moist 40°F followed by 70°F in the light; no germination occurred at 70°F in the light without a temperature cycle. No germination was observed after temperature cycles in the dark, but when the seeds were shifted to 70°F in the light, 90–100% germinated in 3–5 days. The half-life of the seeds was found to be 18 months.

4.3 **Bidens aristosa* (Asteraceae Family), Swamp Marigold or Tickseed-Sunflower

Bidens aristosa, swamp marigold or tickseed-sunflower (height, 0.3–1.5 m), is an annual species that thrives in wet to wet-mesic soils. Swamp marigold is very easy to establish, providing an intense yellow bloom in moist areas from mid-August to early September (illustration on page 158). Because it is an annual that successfully germinates in the spring after a fall planting, it is a good plant to obtain quick results in a new restoration project. It competes moderately well in early-established ecosystems. Consequently, it has a low Coefficient of Conservatism of only 1. Currently, there are no varieties recognized for *B. aristosa*; however, I have grown a formerly recognized variety, *mutica*, which has drastically reduced seed barbs; see seeds scanned on the Springer web site. The northern limit of *B. aristosa* ranges from Maine west to Ontario, Minnesota, Nebraska, and Colorado. From the north it is found south to the states of the U.S. border, except New Mexico and Florida. Immature plants have a leaf structure superficially resembling ragweed, *Ambrosia artemisiifolia*. Seeds can be harvested from mid- to late September after the seed-heads senesce to brown. Flocks of goldfinch compete for seeds; thus, the seed-heads might be clipped a little early when golden tan in color. After drying, the seeds can be jarred loose by scraping over quarter-inch hardware cloth. For germination seeds should be cool-moist stratified for 2 months [11]. As indicated above, a fall planting was found to be completely satisfactory by this author, especially in wet areas of low competition. Most *Bidens* species are colloquially known as "beggar-ticks," "stick-tights," or "Spanish needles" because of the barbed seed's ability to stick to clothing, except for the so-called *B. aristosa* v. *mutica* seeds mentioned above.

4.4 **Calamagrostis canadensis* (Poaceae Family), Bluejoint Grass

Calamagrostis canadensis, bluejoint grass (height, 0.5–1.5 m), is found in wet to mesic prairies, usually in dense continuous patches, which are caused by rhizome spread (illustration on page 158). Bluejoint grass comprises three varieties with somewhat different ranges, but all three are restricted mainly to the northern areas. The typical variety, *canadensis*, ranges over most of Canada and Alaska; in the U.S. it is spread from Maine west to Washington and south, with the exception of the Mid-South and southeastern states. Because seed-heads are sparse in these dense areas of growth, it is likely that propagation of bluejoint grass occurs primarily by root rhizomes. Although the Coefficient of Conservatism is only 3, this grass does require wet to wet-mesic (rarely mesic) conditions. In view of its low level of conservatism and propagation by root rhizome, it may be easier to restore this plant by transplantation of a small sample. Alternatively, seeds can be collected from the sparse number of seed-heads. Seeds require no treatment, but they should be sowed on top of the soil [11]. Because of its ability to spread, the restoration site ideally would have a wet area of relatively small size so that this grass would not become dominant.

4.5 **Carex lasiocarpa* v. *americana* (Cyperaceae Family), Hairy-Fruited/Narrow-Leaved Woolly Sedge

As the name implies, *Carex lasiocarpa* v. *americana*, hairy-fruited or narrow-leaved wooly sedge (height, up to 1.2 m), is uniquely supplied with a dense hairy or woolly perigynia (the sheath surrounding the achene or seed). Although this *Carex* has a wide range, it is fairly uncommon, and perhaps as a consequence it has a Coefficient of Conservatism of 10 (illustration on page 158). This sedge occurs in most of Canada and Alaska in wet to wet-mesic conditions (rarely mesic). In the East it has been reported in Maryland south to Virginia. In the Midwest it ranges from Ohio west to Iowa and Nebraska and all states north of this line. In the West it ranges from Colorado west to California and north, except Nevada. Seed-heads can be harvested from mid- to late June. Seeds can be obtained by procedures described for other *Carex* species (see Chap. 2). Germination techniques are unknown, but techniques recommended for other *Carex* species are probably satisfactory; that is, a 2-month cool-moist stratification and sowing seed on top of the soil. A fall sowing should also be satisfactory.

Another similar more common species having hairy perigynia, **Carex pellita/ lanuginosa*, can be characterized as possessing wider leaves (up to 5 mm), whereas *C. lasiocarpa* v. *americana* has more narrow leaves (up to 2 mm wide).

4.6 **Carex vesicaria* v. *monile* (Cyperaceae Family), Blister or Inflated Sedge

Carex vesicaria v. *monile*, blister or inflated sedge (height, 0.3–1 m), flowers in mid-May. It prefers full sun in wet to wet-mesic soils (rarely mesic). It is often found in dense stands, showing that it spreads by rhizomes. See illustration on page 158. Surprisingly, this plant does have a high Coefficient of Conservatism of 9. This variety of *Carex* has an occasional distribution in the northern half of Illinois [17]. It is largely a northern variety, found in Canada from Labrador west to Manitoba and a southern limit in the U.S. of North Carolina, Kentucky, and Missouri. There are five other varieties of *C. vesicaria* found either in the Northeast and Northeast Canada, or the West and West Canada. Blister sedge seeds can be collected from mid- to late June. The recommended seed treatment is 2 months of cool-moist stratification followed by sowing on top of the soil [11]. Drying of the soil should be prevented as described for *Carex brevior* (see Chap. 2).

4.7 **Eleocharis obtusa* (Cyperaceae Family), Spike Rush

Eleocharis obtusa, spike rush (height, 0.3 m), is a delicate-leaved wetland rush (illustration on page 158). This spike rush is 1 of about 15 *Eleocharis* species found in the tallgrass area, and this particular species is common in the East and Midwest. *E. obtusa* is found in most of southern Canada and the U.S., except the states located in the Rocky Mountains. Very fine seeds are collected in late June. Its habitat is low wet areas, such as along ditches, rivers, and lakes. This is a plant restricted to wet and wet-mesic areas, and in such conditions the delicate aspect of the plant lends a welcome addition to diversity. This spike rush often volunteers in fairly disturbed situations, so it is hardly surprising that it has a Coefficient of Conservatism of only 2. For germination seeds require 2 months of cool-moist stratification and sowing on top of the soil [11].

4.8 **Eupatorium perfoliatum* v. *perfoliatum* (Asteraceae Family), Common Boneset

Eupatorium perfoliatum v. *perfoliatum*, common boneset (height, 0.4–1.5 m), is primarily a wetland plant. As the scientific name *perfoliatum* implies, the leaf circumvents the stem (see illustration on page 159). Thus, the leaf structure is a major feature of interest. The small white flowers, blooming in early September, are not exceptionally showy. Although the typical variety *perfoliatum* enjoys wide distribution in all states south of its northern limit of New Brunswick west to Manitoba and North Dakota, it is mainly restricted to wet or wet-mesic prairies and damp partial shaded areas. One other variety has very limited distribution in New England and Ontario. Occasionally, common boneset will be seen in mesic soils, but it rarely persists there. Its Coefficient of Conservatism of 4 is relatively low, but it does have a fairly strict requirement for wet situations. The seed-head can be harvested in late August to mid-October when a mature fluffy white. Seeds can be easily thrashed loose when the seed-head is dried. The black oblong seed (about 1–2 mm long) is attached to a fluffy white pappus 2–3 mm long. For germination, the seeds need no treatment but require sowing on top of the soil [11]. Experimental study showed that seeds dry stored for 6 months germinated 100% at 70°F in 8–13 days, and light was an absolute requirement for germination [12]. Seeds dry stored for 1 year were all dead [12].

4.9 **Eupatoriadelphus maculatus* v. *maculatus*, Spotted Joe Pye Weed, see *Eupatorium purpureum* (Chapter 2)

4.10 **Filipendula rubra* (Rosaceae Family), Queen-of-the-Prairie

Filipendula rubra, queen-of-the-prairie (height, 0.6–2 m), is true to its name. Its large inflorescence of many pink flowers in late June to early July is striking (see illustration on page 159). This plant is listed as threatened in Illinois, and it rates the highest Coefficient of Conservatism. Its northern limit extends west from Newfoundland to Ontario and Minnesota. The southern limit extends west from North Carolina, Virginia, West Virginia, and north of the Ohio River to Missouri. Its growth requirement is fairly restrictive, as it prefers full sun in wet to wet-mesic soils. It thrives in wet-mesic black prairie silt-loam, and it holds its own in mesic soil provided there is some partial shade. Interestingly, Master Gardeners at King's Gap Environmental Center in Pennsylvania have succeeded in growing this species nearly on the summit of a mountain underlain with sandy soil. According to one Master Gardener, there is a seep in its location. Seedpods (5–8 mm long) can be collected from the seed-head from early to late September. A 3-month cool-moist stratification of seeds is required for germination [11]. This author has found it difficult to restore queen-of-the-prairie from seed. This plant spreads easily by root rhizome. It is relatively easy to restore this plant by transplanting a sample from an existing cluster of plants. However, such a rare plant should not be taken from any site, except those threatened with eminent development. Also, live plants can be obtained from certain nurseries. If these plantings are successful, spreading colonies can be used to transplant into future restoration sites. If your site has a wet area, it is a highly recommended addition to the restoration. This plant readily responds to wet or wet-mesic black prairie-loam. In wettish gravelly clay it suffers, possibly because of periods of summer drying or lack of nutrients.

4.11 **Heracleum maximum, Heracleum lanatum* (Apiaceae Family), Cow Parsnip or Masterwort

Heracleum maximum, cow parsnip or masterwort (height, 1–2.8 m), is widely distributed from the East to the Midwest, being restricted to moist creek bottoms in partial shade or full sunlight (illustration on page 159). It is found in most of Canada and the U.S. However, it is absent from much of the South from Florida west to Texas and Oklahoma. Distinguishing features are the huge wooly leaves divided into three leaflets (0.15–0.6 m long and broad) and large white inflorescences that appear in early June. This plant is an interesting addition to bottomlands and other moist ecosystems. The Coefficient of Conservatism of 6 indicates a moderate tolerance of disturbance. Seed can be readily stripped from the tan seed-head in late June to mid-July. Seed should be planted in the fall, or given alternating cool-moist, warm-moist, cool-moist treatments [11]. In my hands, cow parsnip was difficult to restore from seed.

4.12 ******Hypericum ascyron, Hypericum pyramidatum*** (Clusiaseae Family), Great or Giant St. John's-Wort**

There are more than a dozen *Hypericum* species that can be found in the prairie, but few are more spectacular than *Hypericum ascyron*, great St. John's-wort (height, 0.6–2 m). As shown on page 159, large yellow flowers bloom in mid-July to early August. Each individual flower thrives only a couple of days, and the peak bloom of individual flowers survive only 1 day. However, flowering is prolific with many blossoms continuing for some time. In full sun the petals tend to curl, diminishing the flower's full beauty. Great St. John's-wort thrives in wet-mesic bottomland prairies, but it also can be found in mesic soils in partial shade. This plant is found in Quebec and Ontario in the north, and it occurs in the northeastern quarter of the U.S. It ranges as far west as Nebraska, and south to Missouri and West Virginia. Seeds are borne in relatively hard conical/ovoid capsules (2–3 cm long). Crushing the capsule gives seeds about 1 mm long with a longitudinal wing. Seeds can be sieved through window screen for purification. Germination requires a 2-month cool-moist stratification followed by sowing on top of the soil [11]. With a Coefficient of Conservatism of 8, great St. John's-wort is considered difficult to restore. Under cultivation, great St. John's-wort responded well to fall sowing followed by weeding and regular mist irrigation.

4.13 ******Iris virginica*** v. ***shrevei*** (Iridaceae Family), Southern Blue Flag**

Iris virginica v. *shrevei*, southern blue flag (height, 0.6–0.8 m), is an attractive wetland iris giving blue-yellow flowers in mid- to late June (illustration on page 159). Its preferred ecosystem is wet to wet-mesic soils in full sun or partial shade. It can be grown in mesic soil if competition from other plants is kept low by weeding, but once established under mesic conditions it will persist for several years.

Its northern range is New York and Quebec west to Ontario, Minnesota, and Nebraska. It is found south and east of the northern limit, except New Jersey, Georgia, and Florida. One other variety, *virginica*, mainly exists in the Southeast and Mid-South. Although southern blue flag is uncommon, it has a Coefficient of Conservatism of 5. The seed capsules, a three-lobed long cylinder (5–6 cm long), can be collected when the capsule is brown from early to late August or later. The capsule readily breaks open to reveal tightly stacked brown seeds in two rows in each of the three cells. The seeds are shaped from stacking in the capsule. On one side the seeds are shaped a flat half-moon to three-quarter-moon (5–7 mm in diameter). On the other side, the seeds are usually pie shaped (about 3 mm at the widest). For seed germination, an unusually long period of cool-moist stratification using fresh seeds is necessary. Recommended seed treatment is to lightly cover fresh seed with soil in the fall. Or, one can spring plant after 4 months of cool-moist stratification [11].

4.14 **Liatris pycnostachya* v. *pycnostachya* (Asteraceae Family), Gayfeather

In early August, few sights of the prairie can be equaled by large colonies of the purple-spiked *Liatris pycnostachya* v. *pycnostachya*, gayfeather (height, 0.6–1.5 m). See illustration on page 159. The typical variety, *pycnostachya*, is far more common; one other variety is known only in Texas. This prairie plant does best in wet bottomlands in wet to wet-mesic soil, but it will also thrive in mesic black prairie silt-loam. In other mesic soils it is not very persistent. Gayfeather has a superficial resemblance to *Liatris spicata*, but the differences can be discerned by comparison. *L. pycnostachya* v. *pycnostachya* has a more dense arrangement of the florets and leaves. Also, in the case of *L. pycnostachya* the bracts at the base of the floret are pointed and curve downward, compared to the appressed-oblong round-tipped bracts of *L. spicata*. Gayfeather is found along its northern limit from Massachusetts and New York west to North Dakota. From the north it is found south to Texas and Mississippi. It is absent in the Southeast from Maryland and West Virginia south. The Coefficient of Conservatism is rated at 6. Seeds plus bracts can be stripped by hand from the spike when the tan or gray pappus fluffs out from late August to late October. In contrast to the seed drop encountered with *L. aspera*, the seeds of *L. pycnostachya* are not as likely to be lost over time, except in very high winds or later in the season. Seed and pappus resemble those of *L. aspera*, except they are somewhat smaller (seed 4 mm long; pappus 5–6 mm long). Seeds require a 2-month cool-moist stratification for germination [11]. Although it is intensive work, transplanting the root corm is very successful. Rodents will feed on the root corm, but this does not appear to be a major problem in a wet to wet-mesic bottomland habitat.

4.15 **Liatris spicata v. spicata* (Asteraceae Family), Marsh Blazing Star or Button Snakeroot

Liatris spicata v. *spicata*, marsh blazing star (height, 0.6–1.8 m), resembles *L. pycnostachya* (see above) in both habitat and superficial appearance. See illustration on page 159. *L. spicata* v. *spicata* generally has less dense florets and leaf structure, but more certain identification is its appressed oblong round-tipped bracts at the base of the florets. Its northern limit ranges from Massachusetts, New York, and Quebec west to Ontario, Wisconsin, and Missouri. From its northern area it ranges south and east into all states. One other variety, *resinosa*, is found south from Pennsylvania to Florida and Mississippi. Interestingly, the Jennings Environmental Center in western Pennsylvania has a large expanse of *L. spicata* existing in a natural remnant prairie that seems to be composed of mesic soil. The Coefficient of Conservatism is rated at 7. Flowers appear in mid-August, and seeds plus bracts may be stripped by hand from the spike when the tan pappus fluffs out in early to mid-October. Seed plus pappus resembles those of *L. pycnostachya*, except the seed and pappus of *L. spicata* v. *spicata* are often more variable in size (seed, 3–7 mm; pappus, 4–7 mm). Several seed germination methods have been tested, all of which were not affected by light [12]. Both fresh seeds and seeds dry stored for 6 months at 70°F germinated 100% at 70°F (induction time of 7 days at 5%/day for fresh seeds and 2.5%/day for dry-stored seeds). Three months of cool-moist treatment (40°F) resulted in 60% germination followed by 40% germination when the temperature was raised to 70°F. Thus, fall and spring planting are options. Transplanting root corms has been found to be successful.

4.16 **Lobelia cardinalis* (Campanulaceae Family), Cardinal Flower

Lobelia cardinalis, cardinal flower (height, 0.4–1.7 m) is remarkable by its bright red flowers, an unusual sight in prairies (illustration on page 159). Cardinal flower is restricted to wet or wet-mesic bottomlands, and it is often found along water-filled ditches. The plant seems to prefer partial shade rather than full sunlight. Cardinal flower has a surprisingly large distribution in all states south and east of its northern extent from Maine and Quebec west to Ontario, Minnesota, Nebraska, Colorado, Utah, Nevada, and California. Its Coefficient of Conservatism is ranked at 6. The brown seed capsules can be collected in early September to late October. Very fine seeds, typical of *Lobelia* species, are easy to collect after the capsule dries and opens. Seeds can be purified by passage through window screen. To germinate, seeds require light, but will germinate in the dark in the presence of gibberellic acid-3 [12]. Accordingly, germination in the light at 70°F was 85–90% (induction time, 4–8 days at 2.8–6.5% per day). A 3-month cool-moist treatment (40°F) followed by 70°F in the light gave 90% germination (induction time, 1 day at 15% per day). Seed dry storage has an important effect on germination [12]. When placed in the light at 70°F, the percent germinations for seeds dry stored 0, 6 months, 1 year, and 2 years were 14, 79, 77, and 0%, respectively. Alternatively, seeds could be sowed on top of the soil in either the spring or fall.

4.17 **Lobelia siphilitica* v. *siphilitica* (Campanulaceae Family), Blue Cardinal Flower

Lobelia siphilitica v. *siphilitica*, blue cardinal flower (height, 0.3–1 m), gains its name from its use as an old treatment for syphilis, albeit an ineffective one. It is a bright blue flower that can be found mainly in wet to wet-mesic habitats in full sun or partial shade (illustration on page 160). The variety *siphilitica* has a northern range from Maine west to Ontario and Minnesota. It is found in states south and east of the northern limit, except for Louisiana and Florida. One other more western variety, *ludoviciana*, ranges from the Midwest to the Rocky Mountains. Despite its usual growth habit in wet areas, Norman Deno of State College, Pennsylvania, grows this plant on top of a mesic hillside in partial shade. According to him (personal communication), the plant requires disturbed soil free of competition from other plants; thus, this requirement would probably account for its presence in silt-laden floodplains. Flowers appear in late summer to early autumn. Its Coefficient of Conservatism is 4. The seed capsule is cup shaped to spherical (6–9 mm long; 6–10 mm in diameter), yielding brown elliptical seeds (0.6 mm long). Harvest can be completed after the capsules turn tan in color from early September to early November. Seeds can be recovered by first scraping across quarter-inch hardware cloth followed by sieving through window screen. As shown by Deno [12], seeds require light for germination. Deno also found that germination will occur in the dark only with gibberellic acid-3 treatment. With seeds dry stored 6 months at 70°F, light germination at 70°F was 75–85% (induction time of 30 days at 2.3% per day). Three months of cool-moist treatment at 40°F followed by 70°F in the light resulted in 80% germination (induction time, 2–4 days at 30% per day). Alternatively, a fall sowing on top of the soil has proved to be effective. Plants often become apparent rather late in the season (about July) as small seedlings, but they will flower the first year if given proper care, including weed removal.

4.18 **Lysimachia quadriflora* (Primulaceae Family), Prairie or Narrow-Leaved Loosestrife

Lysimachia quadriflora, prairie or narrow-leaved loosestrife (height, 0.2–0.9 m), is named *quadriflora* for the four flowers clustered at the ends of the branches (not always a reliable four). See illustration on page 160. Flowers appear in late June. In Canada it is found in Ontario and Manitoba. In the U.S. it occurs mainly in the Dakotas, Oklahoma, Arkansas, Alabama, Georgia, North Carolina, and the U.S.

northeastern quarter, except for much of New England. With a high Coefficient of Conservatism of 8, this is an uncommon, special plant. Its preferred ecosystem is wet to wet-mesic (rarely mesic) soils on prairies. Seed-heads can be collected from late September to mid-October, but the seed capsules persist somewhat longer. The capsules should be crushed to recover the seeds. Seeds can be sieved through window screen. For germination the seeds are treated with a 2-month cool-moist stratification followed by sowing on top of the soil [11].

4.19 ***Lythrum alatum* v. *alatum* (Lythraceae Family), Winged Loosestrife**

Lythrum alatum v. *alatum*, winged loosestrife (height, 0.3–1.2 m), derives its name from the angled, slightly winged stems (illustration on page 160). Winged loosestrife prefers to grow in wet to wet-mesic soil (rarely mesic) in full sun. Its northern limit is Maine and Ontario west to Montana. From its northern region it is found south and east, except New Mexico, Texas, Louisiana, and South Carolina. One other variety is localized in the Southeast and Mid-South. As the Coefficient of Conservatism of 5 indicates, it is relatively easy to restore in its favored ecosystem. Flowers bloom in late June, and seeds can be collected in late August to mid-October. The elongated seed capsules are small (5–6 mm long) furnishing tiny seeds. The senesced (brown) capsules can be stripped from the stalk by hand. The senesced capsules tend to readily detach, so many seed-collecting visits are required for maximum yield. It is hardly worthwhile to separate the seeds from the capsule. For germination the seeds need 2 months of cool-moist stratification followed by sowing on top of the soil [11]. Because *L. alatum* has purple flowers, it would be disastrous to mistake it for the related *Lythrum salicaria* (purple loosestrife), the latter of which is an aggressive alien that decimates native wetland species. Purple loosestrife has alarmed ecologists to the degree that insects predatory on this plant have been imported from Eurasia. Although purple loosestrife tends to be only slightly larger (0.6–1.2 m), it has larger leaves (5–8 cm long and 6–10 mm wide) and larger flowers (12–16 mm broad), whereas winged loosestrife has smaller leaves (1.2–3 cm long and 4–10 mm wide) and smaller flowers (6–10 mm broad). Rather than mistakenly collecting the seeds of *L. salicaria* from the wild, it is better to purchase seeds of *L. alatum* from a reliable nursery. The aggressive tendencies of certain Eurasian plants are legendary.

4.20 ***Napaea dioica (Malvaceae Family), Glade Mallow**

Napaea dioica, glade mallow (height, 1.5–3.0 m), is usually found in ditches or on stream banks, but it will thrive on mesic soil with occasional irrigation. The illustration (page 160) gives a closer view of the flowers in the upper left inset. Its large leaves and height give the plant an attractive aspect. Although it has been assigned a fairly low Coefficient of Conservation of 4, glade mallow is rarely seen. In the wild I have observed glade mallow only twice, once in a roadside ditch in Tazewell County, Illinois, and again in a ditch along a rails-to-trails project in Wisconsin. Consequently, glade mallow has a rather restrictive distribution in a line of states from Pennsylvania west to Minnesota and Iowa. Herbivores, probably deer, eat the leaves and developing flower heads. Flowers appear in late June, and seeds can be harvested from mid-August to mid-September when the capsules turn brownish black. A 2-month cool-moist stratification is recommended for seed germination [11].

4.21 **Scirpus cyperinus* (Cyperaceae Family), Wool Grass

Scirpus cyperinus, wool grass (height, 1–1.5 m), is a rather common species of the *Scirpus* genus found in swamps or wet meadows in the East and Midwest of both Canada and the U.S. See illustration on page 160. This species has a sporadic occurrence in the Great Plains and West, being mainly restricted to the North. This rush requires wet to wet-mesic conditions in full sun or partial shade. It has a median Coefficient of Conservatism of 5. In early July the tiny seeds (up to 1 mm long) are easily collected from the seed-head by shaking them over an open paper grocery sack. The small seeds should be sowed on top of the soil shortly after collection or wait until fall.

A similar species shown in the herbarium scans on the Springer website, **Scirpus pendulus/lineatus*, red bulrush (height, 0.3–1.4 m) has the same ecosystem requirements. *Scirpus pendulus* readily yields pale brown seeds (about 1 mm long) by shaking the seed-heads in early August. Certain authorities consider *S. lineatus*, a species restricted to the Southeast, to be separate from *S. pendulus*. There are about five species of prairie rushes in the *Scirpus* genus and about seven in the related rush genus *Schoenoplectus*.

4.22 **Spartina pectinata* (Poaceae Family), Freshwater Cord Grass

Spartina pectinata, freshwater cord grass (height, 0.6–2 m) is a wetland grass that can grow in standing water, but it can be observed occasionally on mesic soil (illustration on page 160). Cord grass is tolerant of both full sun and partial shade. Because cord grass spreads by rhizomes, it usually can be observed as large patches or colonies. Cord grass has an extensive range from Canada's Newfoundland west to Alberta and the Northwest Territory. In the U.S. it is found in all states, except the Southwest and Southeast. Its Coefficient of Conservatism is 4. The grass blades are sharp enough to cut skin by slicing action. A comb-like seed-head flowers in early to mid-August, and it is ready for clipping when senescent tan, about mid- to late

October. Prominent bracts surround the grain. Although no treatment is recommended to initiate germination [11], the restoration of cord grass can be difficult from seed. Because of its rhizomatous nature, it is far easier to dig clumps and divide them for transplanting. If the area is wet, this grass can be very aggressive. For example, it is recommended for pond perimeters or wet areas, but it will not tend to spread much into drier areas.

4.23 ******Verbena hastata* **v.** *hastata* **(Verbenaceae Family), Blue Vervain**

Verbena hastata v. *hastata*, blue vervain (height, 0.6–2 m), is found in full sun or partial shade on swampy ground, mainly wet to wet-mesic, rarely on mesic soil (illustration on page 160). The typical variety, *hastata*, has a northern range from Nova Scotia west to Saskatchewan. In the U.S. it is found in the north from Maine west to the Dakotas, and ranges south and east to all states, including New Mexico in the Southwest. One other variety, *scabra*, is mainly restricted to the western half of the U.S. According to one authority [13], blue vervain is a perennial, but in my hands it seems to be an annual that is highly successful in reseeding itself. At best it would seem to be a biennial. The spikes of flowers are largely blue, but some forms are purplish blue, rose colored, or white. As its Coefficient of Conservatism is 3, it can be rather weedy in wet areas. From mid-September to late October, the seed-head can be clipped and dried when it turns a senescent brown. The reddish-brown seeds (oblong, 1.9–2.1 mm long) readily drop from the seed-head with little thrashing. Alternatively, a greater yield of seeds is obtained by scraping over hardware cloth, but more extraneous material accumulates. Germination studies [12] showed that a 3-month cool-moist incubation followed by exposure to light and 70°F was required for 100% germination (in 3–5 days). Without the prior cool-moist treatment, no germination occurred at 70°F in the light or dark. Further experiments [12] with seeds dry stored for a year at 70°F demonstrated that between 8 and 12 weeks of a cool-moist period was the minimum required for 100% germination at 70°F in the light; fresh seeds were not tested in this latter work. In my hands, restoration is easy with fall sowing on top of the soil.

Mesic/adaptive species tolerating both wet and wet-mesic soils follow: *Carex vulpinoidea, Elymus virginicus, Euthamia graminifolia, Lilium michiganense, Physostegia parviflora, Physostegia virginiana, Pycnanthemum virginianum,* and *Symphyotrichum novae-angliae.*

Those mesic/adaptive species tolerating wet-mesic soils are *Allium cernuum, Amsonia tabernaemontana, Andropogon gerardii, Apocynum cannabinum, Arnoglossum plantagineum, Asclepias sullivantii, Astragalus canadensis, Baptisia alba macrophylla, Camassia scilloides, Castilleja coccinea, Carex brevior, Carex davisii, Carex shortiana, Chamaecrista fasciculata, Cirsium discolor, Comandra umbellata, Coreopsis tripteris, Desmodium canadense, Dodecatheon meadia, Echinacea purpurea, Elymus villosus, Erygium yuccifolium, Eupatorium purpureum, Fragaria virginiana, Gaura biennis, Gentiana alba, Gentiana andrewsii, Geum triflorum, Helianthus grosseserratus, Helianthus maximiliani, Helianthus tuberosus, Heliopsis helianthoides, Heuchera richardsonii, Hypoxis hirsuta, Iliamna rivularis, Juncus tenuis, Lysimachia ciliata, Lobelia spicata, Monarda fistulosa, Oenothera biennis, Oligoneuron rigidum, Packera paupercula, Panicum virgatum, Pedicularis canadensis, Penstemon digitalis, Phlox pilosa, Polemonium reptans, Polytaenia nuttallii, Pycnanthemum tenuifolium, Pycnanthemum verticillatum pilosum, Rosa arkansana, Rosa carolina, Rosa setigera, Rosa virginiana, Rudbeckia fulgida, Rudbeckia hirta, Rudbeckia subtomentosa, Senna hebecarpa, Senna marilandica, Silene stellata, Silphium integrifolium, Silphium laciniatum, Silphium perfoliatum, Silphium terebinthinaceum, Sisyrinchium angustifolium, Solidago canadensis, Spiranthes cernua, Sporobolus heterolepis, Symphyotrichum laeve, Symphyotrichum pilosum, Teucrium canadense, Thalictrum dasycarpum, Tradescantia ohiensis, Tripsacum dactyloides, Veronicastrum virginicum, Vernonia fasciculata, Viola cucullata/affinis/sororia,* and *Zizia aurea.*

Wet-Mesic to Wet Soil Preferring Species

Angelica atropurpurea

Asclepias incarnata

Bidens aristosa

Calamagrostis Canadensis

Carex lasiocarpa v. americana

Carex vesicaria v. monile
Photo: Mark Berhow

Eleocharis obtusa

Eupatorium perfoliatum

Filipendula rubra

Heracleum maximum

Hypericum ascyron

Iris virginica v. shrevei

Liatris pycnostachya

Liatris spicata

Lobelia cardinalis
Photo: Mark Berhow

Lobelia siphilitica

Lysimachia quadriflora

Lythrum alatum
Photo: Mark Berhow

Napaea dioica

Scirpus cyperinus

Spartina pectinata

Verbena hastata

Part III
Restoration

Chapter 5
Restoration Methods

5.1 Landscaping with Prairie

It is an excellent alternative to enhance your home garden with native prairie wildflowers. All you need is full sun and/or partial sunlight. In urban and suburban settings one must be careful to select the most attractive species. And, it is advisable to plant several well-planned beds of monoculture. In this way you will not attract the ire of your neighbors. Because suburbanites are familiar with monocultured beds of flowers, this point cannot be overestimated. As shown in Fig. 5.1, a prairie wildflower garden can greatly complement your lawn and home. Illustrated are *Echinacea pallida* and later-blooming *Ratibida pinnata* (just in bud stage).

For example, the following plants of intermediate to smaller height, as well as their ease of restoration, are excellent choices to consider; that is, if the soil type permits: *Allium cernuum*, *Amsonia tabernaemontana*, *Anemone canadensis*, *Asclepias tuberosa* (friable soil or sand), *Baptisia australis*, *Baptisia bracteata* v. *leucophaea*, *Chamaecrista fasciculata*, *Coreopsis lanceolata*, *Dodecatheon meadia* (best in partial shade), *E. pallida*, *Echinacea purpurea*, *Eryngium yuccifolium*, *Iris virginica* v. *shrevei* (moist), *Liatris aspera* (hillside), *Liatris pycnostachya* (moist), *Lupinus perennis* (sand), *Monarda fistulosa*, *Opuntia humifusa* (sand), *Penstemon digitalis*, *Penstemon grandiflorus* (sand), *Physostegia virginiana*, *Rudbeckia fulgida*, *R. pinnata*, *Rudbeckia hirta*, *Tradescantia* sp., and *Viola* sp. Attractive plants that are more difficult to grow are *Allium stellatum* (dry-mesic), *Amorpha canescens*, *Gentiana puberulenta*, *Camassia scilloides*, *Heuchera richardsonii*, *Hypoxis hirsuta*, *Lilium michiganense* (ensure no virus-transmitting tiger lily in vicinity), *Lithospermum* sp., *Dalea purpurea*, *Phlox pilosa*, *Polygonatum biflorum* v. *commutatum* (partial shade), *Silene regia*, *Sisyrinchium angustifolium*, and *Sporobolus heterolepis* (attractive grass). If larger plants are desired as a backdrop, a number of suitable specimens are *Angelica atropurpurea* (moist), *Baptisia alba* v. *macrophylla*, *Bidens aristosa* (moist), *Eupatorium purpureum*, *Filipendula rubra* (moist), *Helianthus maximiliani*, *Heliopsis helianthoides*, *Hypericum ascyron* (moist), *Rudbeckia subtomentosa* (moist), *Silphium laciniatum*, *Silphium terebinthinaceum*, *Solidago speciosa*, *Thalictrum dasycarpum* (moist), *Vernonia*

H.W. Gardner, *Tallgrass Prairie Restoration in the Midwestern and Eastern United States: A Hands-On Guide*, DOI 10.1007/978-1-4419-7427-3_5,

Fig. 5.1 *Echinacea pallida* in an Illinois prairie wildflower garden

Fig. 5.2 Replacement of an "English" lawn in Pennsylvania featuring *Ratibida pinnata*, *Monarda fistulosa*, and other species

fasciculata, and *Veronicastrum virginicum.* These wildflowers offer a low-maintenance alternative to caring for a traditional "English" lawn (Fig. 5.2).

For quick results, annual, biennial, Heliopsis, or short-lived Coreopsis perennial forbs Bidens should be considered, such as *Rudbeckia hirta*, *Heliopsis helianthoides*, *Coreopsis lanceolata*, and *Bidens aristosa* (moist soil). Figure 5.3 shows an August flowering of *B. aristosa* after a late fall sowing.

Fig. 5.3 Quick results with the annual, swamp marigold (Baptisia aristosa), by fall sowing in a wet-mesic area

5.2 Planting Strategies

In general, prairie can be planted wherever there is sufficient rainfall and full sunlight or partial shade for certain species. This author has cultivated prairie species in both central Illinois and south-central Pennsylvania. Familiarity with the characteristics of the planned restoration site is essential for success; that is, soil type, moisture, and amount of sunlight are most important. Planting seed of the incorrect ecotype in the wrong location is a waste of time, seed, and other resources; failure will be guaranteed.

Soil moisture is one of the most important considerations. "Wet" prairie is defined as standing water or water-saturated soil. Soil types then progress toward drier soil through "wet-mesic," "mesic" (medium soil moisture), "dry-mesic," and "dry." "Dry" soil usually is composed of well-drained sand or gravel, but not necessarily. There are certain "dry-loving" species that have a specific requirement for friable soils rather than strict dry requirements. Friable soils include sand, gravel, and black prairie silt-loam. The effect of soils on species distribution can be mysterious at times. According to their preference for soil moisture, some of the more common prairie species are listed in Table 5.1.

Also included in Table 5.1 are ten indicator species given by Curtis [26] for prairies of different soil moistures in Wisconsin. Curtis's list is slightly skewed toward the wet ratings, which may be the result of the more northerly latitude of Wisconsin. Appendix 1 gives a more complete species listing of moisture preferences, and additionally indicates if soil friability is preferred (sand, gravel, or loam).

Table 5.1 Common indicator species associated with different soil moisture[a]

Dry	Dry-mesic	Mesic
Antennaria neglecta	*Amorpha canescens*	*Allium canadense*
Arenaria stricta[d] (*Minuartia michauxii*)	*Anemone virginiana*[dm]	*Andropogon gerardii*
	Asclepias tuberosa	*Apocynum cannabinum*
Artemisia caudata[d]	*Asclepias verticillata*[dm]	*Asclepias purpurascens*
Artemisia ludoviciana	*Lespedeza capitata*	*Viola sororia*
Asclepias viridiflora	*Baptisia bracteata leucophaea*	*Astragalus canadensis*
Viola pedata		*Symphyotrichum laeve*[m]
Bouteloua curtipendula[d]	*Carex bicknellii*	*Arnoglossum atriplicifolium*
Brickellia eupatoroides	*Carex pensylvanica*	*Carex blanda*
Carex muhlenbergii	*Chamaecrista fasciculata*	*Carex brevior*
Dalea purpurea[d]	*Ceanothus americanus*[m]	*Carex gravida*
Linum sulcatum[dm]	*Coreopsis palmata*	*Comandra umbellata*
Monarda punctata	*Dalea candida*[dm]	*Coreopsis tripteris*
Nothocalais cuspidata	*Dichanthelium oligosanthes*[dm]	*Desmodium canadense*[wm]
Oligoneuron album[d]	*Liatris aspera*[m]	*Desmodium illinoense*[m]
Panicum perlongum[d] (*D. linearfolium*)	*Lithospermum canescens*	*Dodecatheon meadia*[wm]
	Echinacea pallida	*Eryngium yuccifolium*[m]
Penstemon grandiflorus	*Eupatorium altissimum*	*Helianthus grosseserratus*[wm]
Phlox bifida	*Euphorbia corollata*	*Helianthus laetiflorus*[m]
Pulsatilla patens[d]	*Gentiana puberulenta*	*Heliopsis helianthoides*
Solidago nemoralis[d]	*Helianthus mollis*	*Lobelia spicata*
Schizachyrium scoparium[d]	*Helianthus occidentalis*[dm]	*Oenthera biennis*
Symphyotrichum oblongifolium	*Hesperostipa spartea*[dm]	*Panicum leibergii*[m] (*D. leibergii*)
Symphyotrichum sericeum[d]	*Oligoneuron rigidum*	*Parthenium integrifolium*
Verbena stricta	*Ruellia humilis*	*Pycnanthemum tenuifolium*
	Schizachyrium scoparium	*Pycnanthemum* vert. *pilosum*
	Scutellaria parvula[dm]	*Ratibida pinnata*[m]
	Solidago speciosa	*Rosa setigera*
	Sorghastrum nutans	*Silene stellata*
	Sporobolus compositus	*Silphium integrifolium*
	Symphyotrichum oolentangiense	*Sisyrinchium albidum*
		Solidago canadensis
	Symphyotrichum ericoides	*Solidago missouriensis*[m]
	Viola pedatifida	*Sporobolus heterolepis*[dm]

Wet-mesic	Wet	Adaptive (dry-mesic to wet-mesic)
Anemone canadensis	*Calamagrostis canadensis*[w]	*Baptisia alba macrophylla*
Angelica atropurpurea	*Eupatoridelphus maculatus*	*Gaura biennis*
Asclepias sullivantii	*Eupatorium perfoliatum*	*Heuchera richardsonii*[wm]
Camassia scilloides	*Hibiscus laevis*	*Juncus tenuis*
Carex vulpinoidea	*Lobelia cardinalis*	*Panicum virgatum*
Cicuta maculata[wm]	*Lythrum alatum*	*Penstemon digitalis*
Filipendula rubra	*Oxypolis rigidor*[w]	*Rudbeckia hirta*[wm]
Galium boreale[wm]	*Scirpus cyperinus*	*Silphium laciniatum*
Gentiana alba	*Scirpus pendulus*	*Silphium terebinthinaceum*

(continued)

Table 5.1 (continued)

Wet-mesic	Wet	Adaptive (dry-mesic to wet-mesic)
Lathyrus venosus[wm]	*Spartina pectinata*[w]	*Tradescantia ohiensis*
Liatris pycnostachya	*Verbena hastata*	
Liatris spicata		
Lobelia siphilitica		
Lysimachia ciliata		
Lysimachia quadrifolia		
Phlox pilosa[wm]		
Physostegia virginiana		
Pycnanthemum virginianum[w]		
Rudbeckia subtomentosa		
Silphium perfoliatum		
Symphyotrichum novae-angliae[w]		
Teucrium canadense		
Thalictrum dasycarpum		
Veronicastrum virginicum[w]		
Zizia aurea[w]		

[a]Dry and dry-mesic species are usually associated with well-drained hillsides, sandy, or gravelly soils. The superscripts denote species in each category considered to be indicator species by Curtis [26] of Wisconsin prairies: [d]dry, [dm]dry-mesic, [m]mesic, [wm]wet-mesic, [w]wet

If the restoration site is composed of several different soil types, different seed mixtures will have to be blended and labeled. Custom-made seed mixtures will greatly increase the probability of successful restoration.

Preference for full sun or partial shade is another important factor in restoration. For example, shooting star, *Dodecatheon meadia*, will survive in full sun, but this plant much prefers to grow along woodland edges. Another shooting star species, *Dodecatheon frenchii*, represents the extreme of habitat fussiness, as it is only found along the dripline of overhanging rock formations. Generally, if a species has a preference for partial shade and is seeded into a mixed ecosystem, this species will initially grow throughout the site, including full sunlight. However, as competition with full-sun plants increases, the partial shade plants often appear to "migrate" to the woodland edges and diminish or disappear from the areas of full sun. Such partial shade species are considered to be indigenous to savannas or woodland edges. In Appendix 1 various species are categorized as to their preference of either partial shade or full sun. In formulating seed mixes for partial shade plants, one must also consider their moisture regimen; thus, several mixes may be required.

Tolerance to disturbance is another important consideration. Taft et al. [6] ranked plants according to their sensitivity to soil disturbance with a "Coefficient of Conservatism" at values between 0 (weedy in disturbed situations) and 10 (indicator of a pristine ecosystem lacking disturbance). Plants ranked between 8 and 10 are often especially challenging to establish. The rate of success of establishing such species is low without extra effort. It is my experience that certain highly conservative plants are easier to propagate in poor soil. That is, there is a minimum of weed competition in such areas. And, the seed of conservative plants can be scattered in established prairies provided that there is not too much competition; that is, areas where there exist patches of exposed soil. Alternatively, these species can be hand-weeded, seeded in weed-free soil, or propagated in pots, all of which are very tedious procedures. Procedures for obtaining weed-free soil are outlined in the Sects. 5.7 and 5.8. Other plants may require unique procedures. For example, in my hands *Gentiana puberulenta* responded very well to late fall scattering of fresh seed in an established prairie, especially in areas of low competition.

At the Normal, Illinois, Prairie Symposium (March 27, 2009), Bill Kleiman (bkleiman@tnc.org) of The Nature Conservancy's Nachusa Grasslands gave good advice about planning a quality prairie. According to him, the following rules should be followed:

- Make diverse prairie restoration a priority
- Plant a large weight of seed
- Plant a high diversity of seed by starting early
- Plant all species you want on year 100 on year 1
- Do it right the first time
- Better to have 5 fabulous acres than a 50-acre weed patch
- Overseed successful planting with more success
- Gather the space and tools to hold, process, and plant a big seed harvest
- Treat each planting as an experiment
- Solve weed problems first
- Starting over is alright

Although these points are great advice, I would like to add a few comments. Regarding one of my restoration projects, I seeded only tallgrass seed in excess on a winter wheat crop. In subsequent years, forb and shortgrass seed were sowed. For several years only specific forbs became obvious, but after 20 years the prairie came to equilibrium typical of a high-quality prairie; that is, tallgrass diminished to acceptable levels. Of course, a large variety of seeds were overseeded in this project. If you do not have a long time to wait, Bill Kleiman's advice should be heeded. Also, planting a large weight of seed has the added advantage of suppressing the growth of alien weeds (see Sect. 5.7).

5.3 Seed Production

Large quantities of seeds can be expensive, and collecting wild seeds may be restricted. As described in this section, large acreage can be restored piecemeal from managed row crops or beds for seed production.

Fig. 5.4 Row-cropped prairie grown as a seed source. (Two plants on left of compass plant are prairie dock-compass plant hybrids)

I have used row crops in mesic soil defined as Keomah silt-loam (nearly level, originally forest soil, somewhat poorly drained, brown friable silt-loam). Parenthetically, I have found that a similar soil, Ipava silt-loam (nearly level, originally prairie soil, somewhat poorly drained, gray-black friable silt-loam) better supports certain species that one would usually find in more sandy soil, such as *Asclepias tuberosa*, *Asclepias sullivantii*, and *Hesperostipa spartea*, which is probably consequent to the very friable nature of the Ipava black soil.

If you choose to cultivate plants in rows, the rows should be at the very least 40 in. apart, and larger plants will require even more width. Rows are convenient to control weeds by regular tilling between rows (Fig. 5.4). Seeds should be sowed in a row with a slight depression to prevent seed migration. It is preferable, but not necessary, to barely cover the seeds with a light cover of sand, except for very small seeds or seeds that require light for germination. You may opt to occasionally fertilize your crop to aid with seed production. Growing row crops has the advantage of easy control of weeds and facile collection of seed-heads of a single species. With certain species the method has several disadvantages. Some species tend to migrate out of the row by root rhizomes; others spread out of the row by seed drop. Additional difficulties arise simply because of the character of the plant. Limited success with short-lived perennials, or biennials, or with sand ecotypes is understandably not the result of the row method. Table 5.2 lists the results of growing row crops in the somewhat poorly drained mesic forest soil. Understandably, poor results were obtained with sand-loving plants.

As shown in Fig. 5.5, forb cultivation in wide beds is a good option; however, the beds should be narrow enough to allow you to reach weeds in the middle from either side (about 5–6 ft wide).

Table 5.2 Persistence of species when grown as row crops in mesic soil[a]

Persistent(10–20+ years)	Moderately persistent[b](decrease after 6–10 years)	Short-lived[b]
Amorpha canescens	*Allium canadense*	*Asclepias tuberosa* (s/l)
Amsonia tabernaemontana	*Allium cernuum*	*Symphyotrichum oblongifolium* (d/dm/s)
Symphyotrichum ericoides	*Allium stellatum*	*Baptisia tinctoria* (s/g)
Baptisia alba macrophylla	*Asclepias purpurascens* (savanna)	*Bouteloua curtipendula* (d)
Camassia scilloides	*Symphyotrichum laeve*	*Bouteloua hirsuta* (d)
Coreopsis tripteris	*Symphyotrichum novae-angliae*	*Arnoglossum atriplicifolium* (5 years/savanna)
Eryngium yuccifolium	*Baptisia bracteata leucophaea* (dm)	*Coreopsis lanceolata* (s/g)
Heuchera richardsonii	*Senna marilandica*	*Delphinium carolinianum virescens* (d/dm/s)
Iris virginica shrevei	*Desmodium canadensis*	*Desmanthus illinoensis* (~5 years)
Penstemon digitalis	*Dodecatheon meadia* (savanna)	*Echinacea purpurea* (3–5 years)
Potentilla arguta	*Echinacea pallida* (s/g/dm)	*Eupatorium perfoliatum* (w/wm)
Rudbeckia subtomentosa	*Euphorbia corollata*	*Koeleria macantha* (d/dm/s)
Silphium laciniatum	*Lespedeza capitata*	*Liatris aspera* (s/g)
Silphium terebinthinaceum	*Liatris pycnostachya* (wm)	*Liatris cylindracea* (d/s)
Solidago speciosa	*Brickellia eupatoroides*	*Lythrum alatum* (w/wm)
Veronicastrum virginicum	*Monarda fistulosa*	*Monarda punctata* (b/s)
	Dalea candida (dm)	*Dalea foliosa* (~3–5 years)
	Dalea purpurea (dm)	*Onosmodium bejariense occidentale* (s/g)
	Psoralidium tenuiflorum (s/g)	*Rudbeckia hirta* (b)
	Pycnanthemum tenuifolium	*Hesperostipa spartea* (s/l)
	Pycnanthemum virginianum (wm)	
	Pycnanthemum verticillatum pilosum	
	Ratibida pinnata	
	Ruellia humilis	
	Silene regia	
	Silphium integrifolium	
	Sisyrinchium angustifolium	
	Oligoneuron rigidum	
	Teucrium canadense	
	Viola pedatifida	
	Zizia aptera/aurea	

[a] The mesic soil was Keomah (friable silt-loam), which is defined as brown, nearly level, somewhat poorly drained, forest soil. It is emphasized that the lifetime of plants can often be influenced by the soil ecosystem; that is, different soils can give large differences in plant survival

[b]Possible causes for non-persistence are given in parentheses. *d* dry, *dm* dry-mesic, *w* wet, *wm* wet-mesic, *l* loam, *s* sand, *g* gravel, *b* biennial. All plants listed produced seed at some stage. Not listed were unsuccessful species that perished before setting seeds as follows: *Asclepias viridiflora*, *Bouteloua gracilis*, *Penstemon grandiflorus*, and *Viola pedata*. All the unsuccessful species require both dry/dry-mesic conditions on soil with sandy or gravelly well-drained character

The need to aggressively pull weeds is a serious disadvantage with the bed method, especially in the first year or two before native plants overcome competition

Fig. 5.5 Propagation of prairie forbs in beds of 6-ft breadth

with weeds. Thereafter, weeding becomes minimal. This method is particularly applicable to cultivating either bushy species or those that spread by rhizome, such as *Coreopsis palmata*, *Artemisia ludoviciana*, *Physostegia* sp., *Anemone canadensis*, *Tradescantia bracteata*, *Rosa setigera* (bushy), *Helianthus maximiliani*, or *Helianthus mollis*. The milkweeds, particularly *A. sullivantii*, and to some extent other *Asclepias* species, migrate efficiently by rhizome for surprising distances; thus, it may be difficult to restrict certain milkweeds to a specific area. *Symphyotrichum oolentangiense* is fairly short lived, and the seed is spread by the wind, eventually occupying adjacent beds.

Certain grasses are particularly difficult to grow in rows or beds because of aggressive spread of seed and lodging or drooping of seed-heads, such as *Tripsacum dactyloides*, *Andropogon gerardii*, *Sorghastrum nutans*, *Schizachyrium scoparium*, and *Panicum virgatum*. These grasses should be grown in wide beds well separated from the forbs. As these grasses will thrive largely without help, weeding is not necessary. Generally, 3 years are necessary for the grasses to develop. As an exception, the conservative grass, *Sporobolus heterolepis*, can be difficult to grow at times, but good success has been achieved by growing it in beds of weed-free soil.

5.4 Seed Suppliers

Many commercial native seed companies are available, some of which are listed here.

1. Prairie Moon Nursery (numerous native grasses, forbs, and shrubs in various quantities), 32115 Prairie Lane, Winona, MN 55987-9515, Phone: (866) 417-8156, (507) 452-1362; Fax: (507) 454-5238, E-mail: info@prairiemoon.com, Website: www.prairiemoon.com
2. Prairie Nursery (numerous native grasses and forbs in various quantities), P.O. Box 306, Westfield, WI 53964, Phone: (800) 476-9453 (service/orders), (800) 476-9453 (technical), Fax: (608) 296-2741, E-mail: webcs@prairienursery.com, Website: www.prairienursery.com
3. Missouri Wildflowers Nursery (grasses, forbs, and shrubs), 9814 Pleasant Hill Rd., Jefferson City, MO 65109, Phone: (573) 496-3492, Fax: (573) 496-3003, E-mail: mowildflrs@socket.net, Website: www.mowildflowers.net
4. Ernst Conservation Seeds (native grasses and forbs in pound quantities; beware of "naturalized" alien seed also available), 9006 Mercer Pike, Meadville, PA 16335, Phone: (800) 873-3321; (814) 336-2404; Fax: (814) 336-5191, E-mail: ernstsales@ernstseed.com, Website: www.ernstseed.com

5.5 Seed Collection

In general, seeds ripen about 1 month after flowering. Seeds of most wild plants are difficult to collect. The seeds ripen over a period of time, and they usually readily shatter (drop their seeds). When ready, seed-heads, pods, and associated stems senesce into a tan, brown, gray, or black color. Other indicators of proper collection times are when the seed capsule opens or the seeds are hard and colored tan, brown, or black. Seeds topped with a fluffy pappus should be picked before they are borne on the wind. Many ripe seeds or seed-heads indicate their readiness by offering little resistance to hand picking, compared to unripe seeds. A brown senescent stem bearing the seed-head is a sure sign of seed ripeness. A range of prairie seed harvesting times can be found in Table 5.3 arranged in order of the first date of harvest. Also listed in Table 5.3 are the earliest dates that seeds ripen in a 2-week window (within parentheses) as determined by Pauly and Coyle [27]. It is emphasized that the Pauly and Coyle data do not account for the rather long period of seed ripening for many species. Data without parentheses shows the full range of seed collection times. Because of facile seed shatter of certain species, many trips must be made to the field in order to collect a maximum seed yield. Only a few species retain their seeds into the late fall or winter. Seeds from immature first-year plants often ripen approximately a month later, compared with mature plants. Seed collection times, if known, can also be found in Appendix 1 arranged in alphabetical order of the scientific name.

Unless individual ripe seeds must be collected one pod at a time, such as with *Psoralidium tenuiflora* and *Euphorbia corollata*, it is most convenient to use hand-held

Table 5.3 Range of seed collection times arranged according to the first day of harvest[a]

Pulsatilla patens	(5/15 to 5/31)
Geum triflorum	(5/15 to 5/31)
Salix humilis	5/20 to 5/25
Antennaria neglecta	5/30 (5/15 to 5/31)
Viola cucullata/sororia	5/30 to 6/15
Polemonium reptans	5/31 to 6/15
Antennaria plantaginifolia	(6/1 to 6/15)
Pedicularis canadensis	(6/1 to 6/15)
Lupinus perennis	6/5 to 6/19 (6/16 to 6/30)
Oxalis violacea	6/5 to 6/20
Carex gravida	6/5 to 6/25
Hypoxis hirsuta	6/5 to 7/10
Carex vesicaria	6/10 to 6/30
Camassia scilloides	6/11 to 6/24 (7/1 to 7/15)
Tradescantia bracteata	6/11 to 7/1
Hesperostipa spartea	6/14 to 6/20 (6/16 to 6/30)
Viola pedatifida	June and see below
Castilleja coccinea	(6/16 to 6/30)
Geranium maculatum	(6/16 to 6/30)
Saxifraga pensylvanica	(6/16 to 6/30)
Krigia biflora	(6/16 to 6/30)
Carex brevior	6/17 to 6/29
Dichanthelium oligosanthes	6/18 to 6/28
Carex davisii	6/18 to 6/30
Carex lasiocarpa americana	6/18 to 6/30
Sisyrinchium albidum	6/18 to 7/8
Allium canadense	6/19 (7/1 to 7/15)
Eleocharis obtusa	6/23
Packera pauperculus	6/24 to 7/1 (6/16 to 6/30)
Carex shortiana	6/24 to 7/12
Delphinium carolinianum	6/24 to 7/20
Coreopsis lanceolata	6/24 to 8/10
Carex vulpinoidea	6/25 to 7/15
Koeleria macrantha	6/26 to 7/7
Angelica atropurpurea	6/26 to 7/18 (7/1 to 7/15)
Dodecatheon amethystinum	6/29 to 7/13
Scirpus cyperinus	7/1 to 7/10
Sisyrinchium angustifolium	7/1 to 7/17
Comandra umbellata	(7/1 to 7/15)
Helianthemum bicknellii	(7/1 to 7/15)
Taenidia integerrima	(7/1 to 7/15)
Carex aquatilis/substricta	7/2
Heuchera richardsonii	7/2 to 7/17 (6/16 to 6/30)
Dodecatheon meadia	7/3 to 7/17 (7/1 to 7/15)
Tradescantia ohiensis	7/6 to 7/31 (7/1 to 7/15)
Juncus tenuis	7/13

(continued)

Table 5.3 (continued)

Anemone canadensis	7/13 to 8/20 (8/1 to 8/15)
Helianthemum canadense	(7/16 to 7/31)
Phlox pilosa	(7/16 to 7/31)
Blephilia ciliata	(7/16 to 7/31)
Baptisia australis	7/17 to 8/1
Astragalus canadensis[c]	7/20 to 9/5
Lobelia spicata	7/20 to 10/27
Zizia aurea	7/23 to 8/18
Psoralidium tenuiflorum	7/24 to 8/25
Elymus villosus	7/25 to 8/15
Silene stellata	7/25 to 8/20
Arnoglossum plantagineum	7/26 to 9/10
Polytaenia nuttallii	7/30 to 8/14 (9/1 to 9/15)
Thalictrum dasycarpum	7/31 to 9/14 (8/1 to 8/15)
Desmodium illinoense	(8/1 to 8/15)
Lathyrus venosus	(8/1 to 8/15)
Lithospermum caroliniense	(8/1 to 8/15)
Lithospermum canescens	(8/1 to 8/15)
Hieracium longipilum	(8/1 to 8/15)
Cicuta maculata	(8/1 to 8/15)
Galium boreale	(8/1 to 8/15)
Polygala sanguinea	8/1 to 8/15
Physostegia parviflora	8/1 to 9/20
Euphorbia corollata	8/1 to 9/26
Scirpus pendulus	8/6
Penstemon grandiflorus[c]	8/6 to 8/14 (8/1 to 8/15)
Rudbeckia hirta	8/6 to 8/16 (8/1 to 8/15)
Eupatorium purpureum	8/6 to 9/15
Baptisia bracteata leucophaea	8/8 to 9/1 (9/1 to 9/15)
Iris virginica shrevei	8/8 to 8/21
Potentilla arguta	8/8 to 9/5 (8/16 to 8/31)
Silene regia[c]	8/8 to 9/5
Zizia aptera	8/8 to 9/5 (8/1 to 8/15)
Tripsacum dactyloides	8/8 to 10/3
Lythrum alatum	8/8 to 10/16
Heliopsis helianthoides	8/10 to 10/6 (8/1 to 8/15)
Iliamna rivularis rivularis	8/11 to 9/29
Rosa arkansana/sulffulta	8/12 to 9/14
Orbexilum onobrychis	8/12 to 9/25
Ceanothus americanus	8/12 to 9/8 (9/1 to 9/15)
Silphium perfoliatum	8/14 to 9/15 (9/1 to 9/15)
Ratibida pinnata[c]	8/14 to 9/25 (8/1 to 8/15)
Bouteloua curtipendula	8/14 to 10/3 (9/1 to 9/15)
Sporobolus heterolepis	8/14 to 10/8 (8/16 to 8/31)
Mimosa microphylla	8/14 to 10/13
Viola pedatifida	8/14 to 10/23
Polygonatum biflorum commut.	(8/16 to 8/31)

(continued)

Table 5.3 (continued)

Elymus virginicus	8/18 to 9/10
Napaea dioica	8/18 to 9/14 (9/1 to 9/15)
Asclepias tuberosa[c]	8/18 to 9/15 (9/16 to 9/30)
Lilium michiganense	8/19 to 9/2
Baptisia tinctoria	8/19 to 9/10
Silphium integrifolium	8/21 to 9/18 (9/1 to 9/15)
Dalea purpurea	8/21 to 10/8 (8/16 to 8/31)
Dalea foliosa	8/21 to 10/17
Parthenium integrifolium	8/21 to 10/17 (8/16 to 8/31)
Mirabilis nyctaginea	8/24 to 9/12
Pycnanthemum tenuifolium	8/24 to 10/30
Elymus canadensis	8/25 to 10/17 (9/1 to 9/15)
Asclepias viridiflora[c]	8/25 to 10/15
Rosa carolina[c]	8/25 to 10/23 (9/16 to 9/30)
Asclepias sullivantii[c]	8/26 to 10/17
Baptisia alba macrophylla[c]	8/27 to 9/25 (8/16 to 8/31)
Echinacea purpurea	8/27 to 10/8
Liatris pycnostachya	8/27 to 10/25 (9/1 to 9/15)
Echinacea pallida[c]	8/27 to 10/30 (9/1 to 9/15)
Eupatorium perfoliatum	8/28 to 11/5 (9/1 to 9/15)
Panicum virgatum	8/28 to 10/22 (8/16 to 8/31)
Desmanthus illinoensis	8/29 to 10/17
Brickellia eupatoroides	8/30 to 10/15 (9/1 to 9/15)
Eupatorium altissimum	9/1 to 10/25
Helianthus grosseserratus	9/1 to 9/30 (9/16 to 9/30)
Tephrosia virginiana	9/1 to 9/20 (9/1 to 9/15)
Allium cernuum	9/1 to 9/25 (8/16 to 8/31)
Lespedeza capitata[c]	9/1 to 10/17 (9/1 to 9/15)
Agastache foeniculum	(9/1 to 9/15)
Lactuca canadensis	(9/1 to 9/15)
Liatris squarrosa	9/1 to 10/6
Lobelia cardinalis	9/1 to 10/20
Lobelia siphilitica	9/1 to 11/3 (9/16 to 9/30)
Opuntia humifusa[c]	9/3 (8/1 to 8/15)
Rudbeckia fulgida	9/3 to 10/15
Penstemon digitalis[c]	9/4 to 10/3 (9/1 to 9/15)
Amsonia tabernaemontana	9/4 to 10/3
Dalea candida	9/4 to 10/15 (8/16 to 8/31)
Anemone virginiana[c]	9/4 to 10/22 (9/16 to 9/30)
Amorpha canescens[c]	9/5 to 10/17 (8/1 to 8/15)
Desmodium canadense	9/5 to 9/18 (9/16 to 9/30)
Filipendula rubra	9/5 to 9/26
Eupatoriadelphus maculatus	9/5 to 10/2 (9/1 to 9/15)
Monarda punctata	9/5 to 10/16 (9/16 to 9/30)
Onosmodium bejariense occidentalis	9/5 to 10/18
Silphium terebinthinaceum	9/5 to 10/23 (9/16 to 9/30)
Pycnanthemum virginianum	9/5 to 10/31 (9/1 to 9/15)
Helianthus laetriflorus	9/7 to 10/1

(continued)

Table 5.3 (continued)

Teucrium canadense	9/7 to 10/15
Asclepias purpurascens[c]	9/9 to 9/18
Chamaecrista fasciculata	9/11 to 10/2 (9/1 to 9/15)
Helianthus mollis	9/11 to 10/8
Monarda fistulosa	9/11 to 10/8 (8/16 to 8/31)
Liatris aspera	9/12 to 10/23 (9/16 to 9/30)
Verbena hastata	9/12 to 10/25
Gaura biennis	9/13 to 10/2
Hypericum ascyron	9/14 to 10/15
Arnoglossum atriplicifolium	9/15 to 10/3 (9/15 to 9/30)
Hypericum punctatum	9/16 to 10/25
Oligoneuron album	(9/16 to 9/30)
Lilium superbum	(9/16 to 9/30)
Asclepias incarnata[c]	9/16 to 10/1 (9/16 to 9/30)
Phlox glaberrima	(9/16 to 9/30)
Carex bicknellii	(9/16 to 9/30)
Physalis virginiana	(9/16 to 9/30)
Apocynum cannabinum	(9/16 to 9/30)
Vernonia fasciculata[c]	9/16 to 10/16 (9/16 to 9/30)
Solidago juncea	9/17
Helianthus occidentalis	9/17 to 10/30 (9/16 to 9/30)
Bidens aristosa	9/18 to 9/30
Coreopsis palmata[c]	9/18 to 10/3 (9/16 to 9/30)
Allium stellatum	9/18 to 10/3
Silphium laciniatum	9/18 to 10/22 (9/1 to 9/15)
Pycnanthemum vert. *pilosum*	9/18 to 10/31
Rudbeckia subtomentosa	9/18 to 11/8
Spiranthes tuberosa	9/20
Lysimachia quadriflora	9/25 to 10/15
Lysimachia ciliata	9/25 to 10/25
Sporobolus compositus	9/25 to 10/17
Veronicastrum virginicum	9/25 to 10/22 (9/16 to 9/30)
Eryngium yuccifolium	9/25 to 10/29 (9/16 to 9/30)
Liatris cylindracea	9/26 to 10/8 (9/16 to 9/30)
Sorghastrum nutans	9/26 to 10/30 (9/16 to 9/30)
Andropogon gerardii	9/26 to 10/30 (9/16 to 9/30)
Symphyotrichum laeve	9/26 to 11/5
Schizachyrium scoparium	9/27 to 11/11 (8/16 to 8/31)
Ruellia humilis	10/1 to 10/17
Senna marilandica[b]	10/1 to 11/16
Liatris spicata	10/5 to 10/18
Helianthus pauciflorus	10/5 to 10/18
Coreopsis tripteris	10/8 to 10/23
Oligoneuron rigidum	10/8 to 10/23
Gentiana andrewsii	10/8 to 10/30 (9/16 to 9/30)
Gentiana alba	10/8 to 11/13
Physostegia virginiana	10/8 to 11/18

(continued)

Table 5.3 (continued)

Tridens flavus	10/10 to 10/25
Spiranthes cernua	10/10 to 10/25
Symphyotrichum sericeum	10/13 to 11/15
Gentiana puberulenta	10/15 to 11/6
Solidago speciosa	10/16 to 11/29
Spartina pectinata[c]	10/18 to 10/25 (8/16 to 8/31)
Aureolaria grandiflora	10/18 to 11/1
Verbena stricta	10/20 to 11/10
Helianthus tuberosus	10/21
Helianthus maximiliani	10/21 to 10/30
Rosa virginiana[b]	10/22
Solidago canadensis	10/24 to 11/14
Artemisia ludoviciana	10/28 to 11/7
Symphyotrichum oolentangiense	10/28 to 11/22
Solidago nemoralis	10/31
Symphyotrichum amethystinum	11/6
Symphyotrichum oblongifolium	11/8 to 12/5
Symphyotrichum novae-angliae	11/9
Euthamia graminifolia	11/9
Symphyotrichum ericoides	11/11
Apocynum cannabinum[b]	11/21 (9/16 to 9/30)
Oenothera biennis[b]	11/21 (8/16 to 8/31)
Symphyotrichum pilosum	11/21 to 12/2

[a]Harvest dates apply for Hardiness Zones (Agricultural Research Service, USDA) 5 (average minimum temperatures of –20°F to –10°F) and 6 (–10°F to 0°F). Parentheses indicate *earliest* seed-ripening dates according to Pauly and Coyle [27]. The dates without parentheses show the full range of collection dates. The dates within parentheses [27] seem to be not optimal for the following species; that is, too late for *Camassia scilloides* and *Baptisia bracteata* v. *leucophaea*, and too early for *Schizachyrium scoparium*. Other large discrepancies in dates are due to the relative persistence of seeds, permitting a later collection as indicated by dates not enclosed in parentheses. Plants that produce seed after the first year of growth from seed usually produce ripe seeds a full month later than normal

[b]Seedpods or seed-heads are persistent. It is an option to collect earlier/later after pods are senescent

[c]Seedpods have intermediate persistence

clippers or shears to recover entire seed-heads. Wear heavy gloves as fingers sometimes accidentally get in the way. Seed-heads are conveniently stored in paper grocery sacks to permit drying before mechanical seed removal.

For personal use seed does not have to be absolutely pure; thus, separation of seed from the seed-head can employ Third World methods. While processing seeds, wearing a dust mask is strongly advised. Except for a few very large seeds, stems and seeds can be separated using a sieve of quarter-inch hardware cloth. The hardware cloth is attached to a wooden frame about 30 in. × 30 in. square and 10 in. in height.

Fig. 5.6 Sieves of quarter-inch hardware cloth and ordinary window screen (Photo: Bryce Gardner)

Four legs of convenient length or simply elevated by 2×4's can be used to support the wooden-framed sieves. The quarter-inch hardware cloth is also convenient for scraping seeds loose from seed-heads. Similar sieves of window screen are excellent for final separation of small seed, such as seeds of *Potentilla arguta*, *Penstemon digitalis*, *Polemonium* reptans, and species of *Carex*, *Gentiana*, *Artemisia*, *Hypericum*, *Lobelia*, *Monarda*, *Pycnanthemum*, *Sisyrinchium*, and *Silene* (Fig. 5.6). If desired, fines can be removed from larger seeds by the screen sieve. More sophisticated sieving systems can be obtained from seed-cleaning suppliers, such as given in Appendix E of Packard and Mutel [10], but it is unnecessary for the purposes of successful restoration. Intractable seed-heads may require more strenuous methods of releasing the seed. Moderate pounding with the top end of a sledgehammer followed by shuffling and twisting underfoot may be required for species such as *Echinacea pallida* and *Ratibida pinnata*. Intractable seed-heads become more friable as the seed-heads age in the field; however, aging causes some losses to bird feeding. For specific suggestions regarding seed purification, refer to discussions of individual species given above in the List of Species sections. Because the extraneous material remaining after seed purification often has some remaining seeds, one can optionally save this trash for later distribution over the restoration site.

If large quantities of seed-heads are obtained, a chipper or leaf mulcher is a convenient method of collecting seeds plus extraneous material (Fig. 5.7). In the case of possessing a single species by a mulcher, Jim Alwill of the Illinois Dept. of Transportation uses a fanning mill to subsequently purify the seeds (personal communication). If seed-heads of several species are processed by a mulcher, one should take into account their requirements for soil type or soil moisture; that is, keep ecotypes separated.

Fig. 5.7 The "Prairie Dawgs" process seed-heads through mulchers. Note the wise use of a dust mask (Photo: Jim Alwill)

5.6 Seed Treatment

This section summarizes the methods used to stimulate germination of seeds. Also, seeds of various species are organized into categories of recommended treatment for ease of reference. For more detailed information about individual species see Part II. If known, seed germination data for other species are summarized in Appendix 1. The seed treatments are often based on the advice of the Prairie Moon Nursery Catalog [11], which I have found to be generally reliable, but other sources can be consulted [3, 10], including my own advice. For experimentally accurate seed germination data on selected species, Deno [12] has been cited. Germination of wild seeds usually requires very specific treatments depending on the species. A minority of seeds from prairie species will germinate without any prior treatment. Usually wild seeds germinate only after specific conditions are met. Deno [12] points out that certain methods are needed to decompose germination inhibitors. Most often, seeds must experience a winter requiring cool-moist conditions (arbitrarily set at 40°F) for a period of time. One practical method to winterize seeds is to simply sow in the late fall. Another method requires storage of the seeds admixed with moist sand, not dripping wet, for 2–3 months in a refrigerator. Other treatments to reduce inhibitors include dry storage for 6 months. A few other species require early spring sowing. Still other seeds must be sown on top of the soil, such as those having an absolute requirement for light or those very small in size. Seeds of legumes require special variable treatments, which often require pod removal, puncture of a hard seed coat by a process known as "scarification," and inoculation with a specific strain of *Rhizobium* microbes. Each of the different seed treatments is discussed in more detail below.

Although a smaller percentage of prairie seeds need no treatment to germinate, most prairie grasses fall into the no-treatment category. In my experience most of these types of grasses germinate satisfactorily in the spring when sowed in the late fall. If grass seeds are stored over a long period of time, it is strongly advised to refrigerate them to prevent loss of viability and seed vigor. Those grasses requiring no seed treatment are *Andropogon* sp., *Schizachyrium scoparium*, *Bouteloua curtipendula*, *Elymus* sp., *Dichanthelium/Panicum* sp., *Sorghastrum nutans*, *Sporobolus* sp., and *Spartina pectinata*. The grasses *Calamagrostis canadensis* and *Koeleria macrantha* need to be placed on top of the soil, and the latter should be sown in the early spring. Among the forbs, no seed treatment is required for *Symphyotrichum laeve*, *Symphyotrichum novae-angliae*, *Fragaria virginiana*, *Brickellia eupatoroides*, *Lilium michiganense*, *Mirabilis nyctaginea*, and *Physostegia virginiana*. *Symphyotrichum oolentangiense* has the additional requirement of sowing in early spring. Because of the small size of seed or requirement for light, the following require placement on top of the soil; otherwise, no treatment is required: *Symphyotrichum ericoides* v. *ericoides*, *Artemisia ludoviciana*, *Heuchera richardsonii*, *Oenothera biennis*, *Pycnanthemum verticillatum* v. *pilosum*, *Pycnanthemum tenuifolium*, *Pycnanthemum virginianum*, *Sisyrinchium angustifolium*, and *Veronicastrum virginicum*.

Seeds of certain species respond to 6 months of dry storage, but otherwise need no treatment, as follows: *Anemone cylindrica*, *Symphyotrichum oblongifolium*, *Coreopsis lanceolata*, *Echinacea purpurea*, and *Penstemon grandiflorus* (sow in early spring for the latter seed). Dry storage followed by sowing on top of the soil is recommended for *Eupatorium perfoliatum*, *Lobelia cardinalis*, *Lobelia siphilitica*, *Monarda* sp., and *Penstemon digitalis*.

According to Prairie Moon Nursery [11], the seeds of several species can be either given no treatment or 2 months cool-moist stratification as follows: *Asclepias purpurascens*, *Asclepias sullivantii*, *Asclepias tuberosa*, *Asclepias verticillata*, *Helianthus* sp., *Ratibida pinnata*, and *Rudbeckia sp.*; however, I have found that seeds in this category germinate satisfactorily with a late fall planting.

The seed of most prairie forbs and a few grasses require a period of moist-cool temperatures (40°F) ranging from 1 week to 4 months. To save labor it is strongly recommended to sow this group of seeds in the late fall. Otherwise, a refrigerator serves to provide cool temperatures, and moisture is usually provided by admixture with moist sand, not dripping wet. The sand/seed mixture should be sacked in a plastic bag. The bag should have a very small opening to prevent anaerobic conditions, but not enough to cause desiccation, which will easily occur in refrigerators. For large quantities of seed Prairie Moon Nursery recommends moistened coarse grade vermiculite; however, for respiratory health one should avoid vermiculite. Large quantities of seed are impossible to store in an ordinary refrigerator.

As an easy alternative, it is highly recommended to plant/sow in the late fall to "overwinter" naturally, especially for seeds needing early spring conditions. I have had some success with adding water to moisten seed for storage outside in shaded garbage cans. Be certain that the seed is not dripping-wet soaked; only slight moisture is required. If stored outside, the seed must be sowed before the weather warms in spring. Rampant growth of mold is another concern with this method. Among the grass seeds, a cool-moist stratification for 2 or 3 months plus sowing on top of the soil is required for *Carex* sp., *Eleocharis obtusa, Juncus tenuis*, and *Scirpus cyperinus*. The grass seeds of *Hesperostipa spartea, Tridens flavus*, and *Tripsacum dactyloides* also require cool-moist conditions for 2–3 months, but it is best to fall-plant these species. A large number of forb seeds need 2–3 months of cool-moist stratification (or fall planting) as follows: *Allium* sp., *Amsonia tabernaemontana, Asclepias viridiflora, Symphyotrichum sericeum, Aureolaria grandiflora* v. *pulchra, Bidens aristosa, Arnoglossum* sp., *Camassia scilloides, Cirsium hillii, Coreopsis palmata, Coreopsis tripteris, Delphinium carolinianum* ssp. *virescens, Echinacea pallida, Eryngium yuccifolium, Eupatorium altissimum, Eupatorium purpureum, Filipendula rubra, Geum triflorum, Iliamna rivularis* v. *rivularis (remota), Liatris* sp., *Napaea dioica, Oxalis violacea, Parthenium integrifolium, Phlox bifida, Physostegia parviflora, Packera paupercula, Silene* sp., *Silphium* sp., *Oligoneuron rigidum* (early spring), *Solidago speciosa* (early spring), *Teucrium canadense, Thalictrum dasycarpum* (early spring), and *Vernonia fasciculata*. Several forb seeds require both 2–3 months cool-moist stratification and sowing on top of the soil (small seed or requirement for light) as follows: *Antennaria* sp., *Asclepias incarnata, Symphyotrichum pilosum, Castilleja coccinea* (also parasitic on grasses), *Gaura biennis, Gentiana* sp. (best fall sown), *Hypericum* sp., *Hypoxis hirsuta, Lysimachia* sp., *Lythrum alatum, Potentilla* sp., *Solidago juncea, Solidago nemoralis, Spiranthes* sp., *Verbena* sp., and *Viola pedata*. Certain forb seeds have optimal time requirements for cool-moist stratification as follows (time in parentheses): *Dodecatheon meadia* (21 days, sow on top of soil in March), *Euphorbia corollata* (30 days), *Iris virginica* v. *shrevei* (120 days), *Pedicularis canadensis* (30 days, parasitic), *Polytaenia nuttallii* (120 days), *Ruellia humilis* (14 weeks), *Tradescantia* sp. (120 days, early spring), and *Zizia* sp. (120 days, early spring).

Certain other forb seeds require more complex treatments. For example, *Ceanothus americanus, Lithospermum* sp., and possibly *Iliamna rivularis* v. *rivularis* (*remota*) should be scarified, as well as cool-moist stratified. Scarification is a process of making a hole in the impervious seed coat by gently rubbing with medium grit sandpaper. As sometimes recommended, boiling water is *not* a replacement for scarification. According to experiments by Deno [12], hot water between 150 and 180°F is fatal to seeds. *Onosmodium bejariense* v. *occidentale, Rosa arkansana*, and *Rosa carolina* seeds not only should be scarified, but also require more complex temperature cycles (see Part II for specifics). Complex temperature cycles are also recommended for *Anemone canadensis, Angelica atropurpurea, Phlox pilosa, Polygala senega, Rosa setigera, Rosa virginiana, Viola* sp. (also known as *papilionacea*), and *Viola pedatifida*. Seeds found in the fruit of *Opuntia humifusa* and *Physalis* species should be removed from the fruit and washed; however, germination of the former seed can be a mystery.

Seeds of legumes generally germinate under different conditions than most other plants. Legumes have seedpods that may or may not be necessary to remove. Large or intractable pods should be removed, such as found in species of *Astragalus*, *Baptisia*, *Chamaecrista*, *Desmanthus*, *Lupinus*, *Psoralidium*, *Orbexilum*, *Mimosa*, and *Tephrosia*. However, seeds are automatically ejected from the drying pods of several species, such as *Chamaecrista fasciculata*, *Chamaecrista nictitans*, *Lupinus perennis*, *Mimosa microphylla*, and *Tephrosia virginiana*; thus, mechanical methods are not required for these species. Other legumes have an inconspicuous hull (soft pod) enclosing the seed, such as species of *Amorpha*, *Desmodium*, *Lespedeza*, and *Dalea*. One can ignore removing the hulls of *Desmodium* and *Lespedeza* species, but *Dalea* species and *Amorpha canescens* are so difficult to restore that it is advisable to remove the hull by rubbing the hulled seed over steel window screen until the seed and hull drop through. Place the window screen an inch or two above a hard surface so as to preserve the integrity of the screen and permit the passage of seed. More vigorous methods may damage the seeds. Certain legumes with a hard seed coat, especially species of *Astragalus*, *Baptisia*, *Chamaecrista*, *Desmanthus illinoensis*, *Lupinus perennis*, *Psoralidium* sp., *Mimosa microphylla*, and *Tephrosia virginiana* may be scarified using fine or medium grit sandpaper; that is, a hole should be made in the seed coat. Failure to scarify usually will delay germination of seeds the following one, two, or more seasons after a late fall planting. For example, a fall planting of *Baptisia alba* v. *macrophylla* and *Baptisia bracteata* v. *leucophaea* results in a percentage of delayed germination until the following year or later. It was estimated that *Baptisia alba* v. *macrophylla* germinated 50% the first summer, 40% the second, and 10% the third. As a consequence, this author rarely scarifies seed unless the seed of the particular species is scarce. Occasionally, a species is encountered that prematurely germinates in the fall, such as *Orbexilum onobrychis*, and the seed of such legume species should be held until spring. After the treatments described above, many legumes also should receive a short 10-day cool-moist stratification, with the exceptions of *Amorpha canescens*, *Desmanthus illinoensis*, *Desmodium* species, and *Dalea* species. Seed suppliers sometimes provide inoculums of specific nitrifying microorganisms to ensure successful growth of legumes. Before sowing, the seeds may be inoculated with a specific strain of *Rhizobium* microbes. In my opinion, inoculation may be most useful with species of *Dalea*, particularly because the tiny fragile seedlings seem to survive more successfully. In my hands other legumes are successfully grown without inoculation of the seeds. As specifically instructed above, seed treatment of the conservative species of *Dalea* and *Amorpha canescens* must be given close attention. Although *Dalea* species and *A. canescens* will grow in rich soil among other prairie plants, starting these plants will require either multiple treatments of the area with glyphosate herbicide or selecting an established prairie ecosystem with minimal competition, often in poor soil. In my opinion it is useful to germinate *A. canescens* in pots for planting out later. If *Dalea* spp. are in short supply, starting these seedlings in pots is also advised.

Despite occasional problems and the exceptions just discussed, the easiest method by far is simply a late fall planting (late October to early November). Many years of sowing seed in the fall have proved the validity of the method with some exceptions.

With most other species, fall planting/sowing avoids tedious procedures involving both specific seed treatment and cold-moist storage. Predation of seed over the winter, low germination rates of certain species, and an early spring flush of weeds are the main disadvantages of the method. Predation can be reduced if the seed is sowed just before the first snow. If the seed does not require treatment (such as certain grasses), a spring or summer planting would be appropriate. However, fall planting of most grasses is also successful in my hands. Of course, proper soil preparation is vital, as discussed in Sect. 5.7.

5.7 Soil Preparation

Alternative for preparation of 1 acre or less. Best results with soil preparation are obtained with glyphosate (isopropylamine salt)[1] herbicide treatment. Monsanto had patent rights to glyphosate with the trade name Round-up, but less expensive glyphosate is now available under other trade names, such as Shoot-out and Ultra-Kill. Many Round-up formulations now include diquat dibromide, which reportedly is persistent for 7 days. A solution of 1% glyphosate herbicide should be used in mid- to late September. After weed kill (2 or 3 weeks later), stray weeds are sprayed again with glyphosate. If weeds are a serious problem, one can start spraying herbicide in early or midsummer, and then repeat whenever weeds start reappearing. Some experts recommend repetitive tilling over an entire growing season to reduce the seed bank, but this method is useful only if rhizomatous plants, such as *Bromus inermis*, have been previously killed with glyphosate herbicide. If not herbicide killed, the chopped rhizomes of *B. inermis* and other perennials will easily reestablish. Other methods of obtaining reasonably weed-free soil are sod removal, steam soil sterilization (requiring expensive specialized equipment), or use of recently exposed subsoil. Shortly before sowing the seed in late October or early November, one pass with a roto-tiller or harrow is useful to prepare the soil. Some experts believe the soil should be well packed. However, it has been my experience with fall sowing that some soil disturbance facilitates slight seed cover with the first rain or snow. Do not cover the seeds with soil. Obviously, small seeds, which are often included in the mix, should not be covered. If the plot is not too large, one can consider packing the seeded soil with a roller or patting the soil with a flat shovel or pressing your weight on a piece of plywood. If you pack the soil, spread small seeds after the packing operation. If the area was contaminated with much weed growth, the seeds can be sowed directly over the thatch left after glyphosate treatment. This latter method is quite effective, and it has the advantage of preventing weed seed being tilled to the surface. Unusually thick thatch may prevent the seed from falling to the surface, in which case burning after herbicide kill is an option. Generally, a heavy planting of native seed will overwhelm germinating alien perennial weeds.

[1] Usually glyphosate isopropylamine salt will be referred to as simply "glyphosate".

See Sect. 5.8 for planting rates. Less hardy annual weeds will disappear after 2 or 3 years, but high mowing of lush weed growth is recommended during the first year (see Sect. 5.8). Be aware that with previous treatment of the soil with persistent herbicides, such as various chlorinated herbicides, herbicide carryover may inhibit restoration of certain species, especially non-grass plants.

Alternatives for planting large acreages. There is some advantage in using an intensively farmed area, as weed invasion is less likely. In soil suspected to contain a large amount of weed seed, such as found in old fields with a large percentage of Eurasian exotics, fall sowing of prairie seed over glyphosate-killed thatch is recommended. Soil disturbance brings weed seeds to the surface where they have an opportunity to germinate. The no-till method is the least work intensive for large acreages. In the case of fields previously infested with weeds, the use of a slightly larger amount of prairie seed could be used.

Large acreages will require farm equipment obtained by contracting with farmers or farm agencies. Glyphosate herbicide treatment should be completed in September. Because only one application of glyphosate herbicide will be practical, a higher than normal concentration of 1.5% should be used. Depending on the spray capabilities of the equipment, the optimal concentrations for total kill of perennials could be discussed with the contractor. Make sure the herbicide covers the entire area. One of my projects has a 10-ft-wide strip of cool-season orchard grass that persists.

Contractors of large prairie restoration projects use range drills to plant seed on tilled soil or herbicide-killed thatch. There are several drills available, such as the Great Plains, Rangeland, Nesbit, Marliss, or Truax drills. Such equipment is out of reach of most amateurs, but contractors with varying capabilities can be found. For example, Peter Schramm of Galesburg, Illinois, has specially adapted a range drill to handle prairie seeds of various sizes. Certain drill settings have some problems with the jamming of fluffy seeds, such as those of *Schizachyrium scoparium*. Also, range drills are not a very satisfactory method for either small seeds or seeds requiring light for germination.

One successful method involving a range drill is certainly worth mentioning. The author contracted to have a field sprayed with glyphosate in September. Late that fall a mixture of forbs seed was hand-sowed in the field without tillage. The following spring a range drill planted prairie grass seeds consisting of a 2/1/1 mix of *Schizachyrium scoparium*, *Andropogon gerardii*, and *Sorghastrum nutan*s. Both plantings were highly successful. Drilling grass in the spring had no notable effect on the growth of the forbs, including species with small seeds. This area now has a surprising concentration of the small-seeded *Potentilla arguta* and *Penstemon digitalis*.

Another successful method involved hand-sowing prairie seed over winter wheat. The following growing season indicated a near-total failure, but this assessment proved to be totally false.

One should be concerned about fencerows or other peripheral areas. Many instances of weed invasion from these untreated areas can be encountered. This author has regretted ignoring the presence of parsnip, crown vetch, teasel, chicory, smooth brome, Japanese honeysuckle, and knapweed along adjacent roadsides and fencerows. After spreading into the restored prairies, action was required to eliminate these pests. Weed spread is less likely into mature prairies, but in newly restored prairies it is advisable to aggressively herbicide the weed-infested peripheries.

Some fields may have residual native prairie, including a percentage of brush and trees. An option is a regimen of burning on a 2-year or annual basis. Some manual cutting and herbiciding of brush and trees may be required. In these cases, it is desirable to upgrade by sowing seed directly on the degraded area. This method has a proven record of success, but it may take as long as 8 years or more before the added species is noticed. I was surprised to see a healthy strip of tallgrass 2 m wide and about 100 m long where seed was sowed many years before in an area previously devoid of tallgrass. By overseeding, a railroad right-of-way was upgraded to Illinois Nature Preserve quality; it now has been dedicated as the Brimfield Railroad Nature Preserve. Keep in mind that Nature Preserve status requires the use of local seed. Research has shown that there are regional genotype differences within the same species. However, as discussed above in the section on *Solidago canadensis*, the genes to adapt to new conditions may be already present, which is a subject worthy of more research.

5.8 Planting Rates

As mentioned above, the seed species mix to sow on the site must match the soil type, particularly the available moisture. You will have to decide on the ratio of grass to forb seed and whether to minimize tallgrass in favor of shortgrass. In my opinion, the use of shortgrass will give quicker results with forbs. Tallgrass can be initially aggressive. Areas planted with a dense initial stand of tallgrass will eventually come to equilibrium with forbs over a decade or more. Prairie Moon Nursery [11] sells mixes of grasses to forbs of 4:1, 3:2, and 1:1. Thompson [28] recommends grasses to forbs mixes of anywhere from 3:2 to 1:2. It is my preference to use a shortgrass to forbs mix of about 1:2, which could include a small amount of tallgrass (about 1–5%). Prairie Moon Nursery [11] recommends 8–10 lbs. per acre of their pure seed mixes. According to Thompson [28], the seed rate should be 14–18 lbs. per acre of pure seed mix. Less purified seed in bulk may require 20–41 lbs. per acre [28]. For guaranteed results, I have used as much as 150 lbs. per acre of crudely purified seeds. Such a heavy planting proved to minimize weed competition in a previously weedy area. On slopes subject to erosion, a cover crop of annual grass should be used. This author has successfully used annual rye and wheat; oat is another suitable cover crop. Thompson [28] suggested annual rye at 1–2 lbs. per acre.

The first year or two after restoring a prairie, expect a profusion of annual and biennial weeds. If the growth of weeds is too lush, your prairie plants may be overwhelmed. When annual weeds reach about 12 in. tall, the plot can be mowed a few times over the summer about 4–6 in. from the ground. It is at this point that some become discouraged enough to abandon their restoration project. Unless some grievous errors have been made, definite evidence of prairie plants will appear within 3 years or, at the latest, 5 years. *Patience is highly recommended.*

Again, it is emphasized that late fall planting is the simplest route to success, but springtime is possible using properly stratified seeds. Of course, many grass species do not require stratification. Stratified seeds should be planted from late May to mid-June. Seeds can be planted as late as mid-July, largely avoiding a flush of annual weeds; however, late plantings usually require irrigation during this normally dry period.

In some areas, establishing conservative species can be a challenge. Peat or plastic pots can be used to start certain conservative species, such as *Amorpha canescens* and *Dalea* spp. before out-planting. A mix of sand and potting soil is satisfactory with a light layer of sand covering the seed. It is advisable to moist incubate peat pots and potting soil months in advance of planting to be certain these materials are depleted of mold nutrients. Molds can be devastating to seedlings. Molding can be avoided with the use of plastic pots and ordinary garden soil with a slight layer of sand to cover the seeds. After being set out, peat pots are targets for digging animals, perhaps squirrels or raccoons, which dig up peat pots to investigate the contents. Chicken-wire cages can be employed. The tedious procedure of growing plants in peat pots is almost an act of desperation, and thus it usually should be avoided with the exception of propagating very conservative species that do not respond to ordinary methods. Once established, conservative plants usually slowly self-propagate.

Part IV
Management

Chapter 6
Prairie Management

6.1 Food for Insects and Wildlife (Plant Predation)

As discussed above, native prairie plants can be a bonanza for insects, herbivores, and secondary feeders. There can be no question that a rich diversity promotes wildlife compared to monoculture agricultural crops or sterile suburban lawns featuring exotic landscaping. Prairie can be a definite benefit for persons who enjoy the presence of wildlife.

However, after prairie "crops" become decimated by insects and wildlife, one can lose some of this positive perspective. Because millions of buffalo once grazed the North American prairies, it is strange to mention plant predation by various animals. However, most present-day herbivores, such as deer, were previously alien in grasslands. This author has witnessed a large population of *Silphium terebinthinaceum* steadily decrease because of the large numbers of deer in a state park. Because deer selectively browse several species, overpopulation can be a serious problem, akin to a herd of goats. For example, deer grazing is particularly detrimental to *Anemone virginiana*. Scattering dog hair around target plants seems to solve the problem. Rodents, such as squirrel and mice, munch on or dig up certain roots, such as those of *Liatris aspera*.

Goldfinch swoop in to harvest seed crops, particularly *Helianthus* species, *Heliopsis helianthoides*, *Liatris aspera*, *Bidens aristosa*, and *Coreopsis* species. Mature seed-heads of *Echinacea* species, *Ratibida pinnata*, *Rudbeckia* species, and *Silphium* species also attract goldfinch feeding. Interestingly, goldfinch flocks are attracted to large concentrations of seed plants versus those isolated in the field. As a result, growing crops for seeds of certain species can be vexing.

Insects are necessary food for birds, amphibians, reptiles, and others. Insects are usually selective about the target species of flowers, seeds, or plants. However, it can be frustrating to find your seed crop of *Gentiana* species being eaten by maggot-like larvae. In the Midwest I have had problems with weevil infestation of the seed of *Baptisia* species, which can be solved with a few days of freezing or insecticide use.

Of course, butterfly larvae feed on specific plant species, such as Monarch larvae on certain species of milkweed, particularly *Asclepias viridiflora* and *Asclepias tuberosa*. Cultivated beds of *Asclepias* species seem to invite much greater infestations

H.W. Gardner, *Tallgrass Prairie Restoration in the Midwestern and Eastern United States: A Hands-On Guide*, DOI 10.1007/978-1-4419-7427-3_6,

of larvae compared to isolated plants in the field. Often the first growing season is relatively free of feeding, but not successive years. Intuitively, it seems impossible, but there seems to be some "memory" for these butterflies to return the next year. For the cause of butterfly conservation, one should tolerate the predatory munching of Monarch larva. Selective tolerance of plant predation will be rewarded by an incredible diversity of wildlife. See Appendix 1 for details of wildlife feeding ("predation"), if known.

6.2 Prairie Ecosystems in Farming Practice

In an age of declining fossil fuel, it makes practical sense to consider low-maintenance prairie perennials in farming methods. Most prairie grasses are not only perennials but are also warm season "C-4" plants. "C-4" plants, such as corn (maize), are highly efficient in utilizing carbon dioxide; thus, they produce record amounts of dry matter in a season. For example, consider the comparatively low yield of dry matter from the soybean, a "C-3" plant.

Farmers or ranchers in certain states in the Midwest have recognized the value of prairie hay as cattle feed. Managed grazing has some value in prairie maintenance. However, it is even less energy intensive to graze our native bison on prairie lands. Bison are best adapted to our climate and native plants requiring less intensive husbandry and manipulation of forage. Lean bison steaks are often preferred to those from the European stock. It is now possible to obtain bison cuts in most areas of the U.S., and the supply is growing.

The harvest of native perennial grains and legume seeds for human food is not even close to being practical, not to mention vegetables and fruit. It will require very intensive breeding programs to increase yield and adapt these plants to agricultural practice, as well as food use. Wes Jackson of the Land Institute in Kansas has been pursuing the goal of breeding a perennial crop from prairie plants, as well as traditional crops. After all, our current supply of plant foods has been derived from thousands of years of human intervention and manipulation. DNA cloning methods may facilitate any potential attempt in adapting prairie plants to human use. In this scenario, there may be an unfortunate loss of native plant stock. South American natives developed several of our more important foods from original native stock, such as potatoes, maize, tomatoes, and the squash family.

This author is skeptical of recent proposals to use "C-4" plants, such as switchgrass, to generate biofuel. The technology for efficient conversion of cellulose to usable fuel may be incredibly difficult and fraught with much loss to entropy (Third Law of Thermodynamics, requiring energy losses to disorder). Even the facile conversion of corn to ethanol yields a negative energy output!

The U.S. Department of Agriculture's Natural Resources Conservation Service (NRCS) has several programs available for soil and water conservation.

The Wildlife Habitat Incentive Program (WHIP) applies to upland, prairie, forest, wetland, and aquatic habitats where native plantings are stressed. For long-term agreements, NRCS can pay up to 90% of costs to install conservation practices. Recent preferences for the WHIP program are aquatic habitat, at-risk species, and reducing invasive species. The Wetland Reserve Program (WRP) restores wetlands at 100% of land value and restoration costs for an easement in perpetuity and 75% of value and costs for a 30-year easement. The Conservation Reserve Program (CRP) requires a contract of 10–30 years for acreage to be planted in "tame" (i.e., alien) grasses, native grasses, wildlife plantings, trees, filter strips, or riparian buffers. The disadvantages of CRP are the lack of requirement for natives and a low annual rental rate, the latter of which discourages farmers. The Conservation Reserve Enhancement Program (CREP) requires contracts of 10–15 years to maintain riparian buffers planted in trees, grass, and other protective vegetation. Mowing is prohibited from April 15 to August 15. The Environmental Quality Incentives Program (EQIP) requires NRCS to develop a conservation plan for landowners that provides payments up to 10 years in duration.

6.3 Global Warming

Carbon dioxide levels today are higher than 800,000 to 2 million years ago. Recent levels are causing huge glacial and polar ice melts worldwide. When meltdown becomes complete, sea levels are estimated to rise about 1 m (about 3 ft). Other than these effects, it is difficult to accurately pinpoint weather changes. Certainly global weather instability is anticipated. Some predict warming, and others even suggest cooling from changes in ocean currents resulting from the open Arctic Ocean. More rainfall will occur in some areas and drought in others.

What are the possible effects of global warming on prairies? As mentioned in the foregoing section, many prairie plants, especially the warm season grasses, employ "C-4 carbon metabolism" for carbon dioxide fixation. As discussed, the C-3 carbon types require higher concentrations of carbon dioxide to thrive, whereas the C-4 carbon types are efficient scavengers of low carbon dioxide concentrations. C-4 carbon metabolism was a strategy that undoubtedly permitted the warm season grasses to predominate in the past. While Earth's carbon dioxide concentration continues to rise, there are potentially serious consequences to perturbing this important ecological variable. One can surmise that C-3 plants will begin to reclaim their competitive edge at the expense of those plants that evolved to take advantage of the lower carbon dioxide concentrations in the pre-industrial age. It seems obvious that Eurasian cool season invaders might become an even more serious problem than the current near-hopeless problem of aggressive alien plants. Predicted changes in rainfall, as well as climate shifts to the north, will aggravate problems for today's prairies.

6.4 Fire Management

6.4.1 *Rationale for Use of Fire*

Because prairie ecology developed over millennia of exposure to fire, regular burning is an important tool to maintain the health of the prairie. With the advent of colonization by the descendents of Europeans, fire was obsessively suppressed, leading to displacement of prairie plants by trees and brush, especially in areas receiving relatively high annual rainfall. Fire is the best method of inhibiting the growth of woody trees and brush by killing the aboveground parts. With some exceptions, woody vegetation will return by sprouting from the root, and persistent prairie fires are required to decimate their prevalence. A hot fire will decimate cool season Eurasian annuals, but biennials and perennials generally recover from the root. Only repeated late burns harm sweet clover (biennial) and smooth/Hungarian brome (aggressive perennial). Burning also returns nutrients to the soil, such as potassium and phosphorus. Additionally, the exposed soil warmed by the sun increases the early growth of prairie plants, as well as beneficial soil microbes. As proof, test plots show much improved growth in burned sections versus unburned. Repeated annual burning are said to favor prairie grasses over forbs; thus, it may be better policy to burn on a 2- or 3-year or more cycle depending on the health of the particular site. With mesic prairies, one cannot wait much longer than 3 years before woody species begin to dominate. In former times, wide expanses of prairie devoid of trees obviously lacked the presence of tree and brush seeds, but ecosystem fragmentation is a serious problem nowadays. Suburbanites typically plant exotic plants and trees, which spread to adjacent land. Ironically, prairies in former times undoubtedly required less frequent fires to prevent the incursion of brush and trees.

Burning suburban prairie gardens will be a difficult sell, if not impossible, in most neighborhoods. An alternative to burning is described below in the Sect. 6.4.7.

6.4.2 *Optimum Times for Burning*

Generally, prairies are ideally burned in March, which blackens and exposes the soil to warming by the sun. In March the prairie has maximum dead/dry duff, which has also been made flammable by the returning warmth of the sun. Early April to early May has been used as an alternative, but at this time early season prairie natives will be inhibited. The late burn method (mid-April to early May) has been used to suppress Eurasian cool season weeds (see Sect. 6.5). Many prairie plants are warm season plants, whereas weedy species from Europe prefer cool seasons. Thus, burning favors warm season plants to the detriment of cool season aliens. However, late fire can be overrated as a method of weed control, except with degraded prairies. This author does not recommend late burns of quality prairies.

Because of residual moisture within plants, late fall burns are often difficult, being restricted to mainly areas of tallgrass. During occasional dry periods, late winter burns can be scheduled, but these burns do not result in maximum killing of exposed portions of woody trees and brush.

6.4.3 *Tools for Fire Control*

On the day of a potentially dangerous prairie fire, there is no substitute for being supplied with the proper firefighting tools. The five essential ingredients are proper initial planning, lighting a back-burning fire (burning into the wind), keeping it under control by swatting, a backup water supply, and communications with other members of the crew. It is essential to have enough fire swatter flaps about three to four to provide a hypothetical crew of six to eight. Fire swatters consist of a 12 in. × 15 in. reinforced rubber flap with a 5-ft handle (see Fig. 6.1). A less

Fig. 6.1 Firefighting tools. *Left to right*: backpack water pump, flat shovel for swatting fire, rubber fire swatter (more efficient), drip torch, garden rake for dragging burning duff (less efficient than drip torch), and side view of backpack water pump

satisfactory alternative for swatting out fires is a flat-bladed shovel. Some prefer firefighting brooms for swatting and sweeping duff into the fire-line during back-burns. As an alternative to fire swatters or brooms, Norm Miller (personal communication) of Brimfield, Illinois, says that using a gasoline-fired leaf blower with another person with a backup leaf blower is an easy way of putting out the windward side of a backfire. Reportedly, the procedure speeds up the progress of setting the backfire. By this method, a backfire can be extinguished at walking pace. According to Todd Remaley (personal communication), the National Park Service also uses leaf blowers to control backfires in certain Western parks. Another important tool is a backpack firefighting pump. There are one- and two-hole nozzle fittings for use as streams of long or shorter distances, respectively. For the most part, the two-hole nozzle is more useful. A handy vinyl collapsible 5-gallon pack is available, but the most reliable pack is a galvanized metal 5-gallon tank (shown in Fig. 5.5). We have found that the vinyl packs crack and develop leaks over time. It is preferable to have at least two backpack water pumps available, so that one can be filled while the other is in action. You can obtain Drench water additive from firefighting supply companies, which improves the performance of water, or alternatively a small amount of liquid detergent can be added to the tank. Although extra 5-gallon water tanks (plastic with handle) are difficult to carry into remote sites, they are often needed, and sometimes they can be replenished on-site. Two drip-torches are handy to encircle a site starting in opposite directions from downwind. Although suppliers recommend the use of only kerosene or heating fuel for safety reasons, this fuel is unsatisfactory in cool weather. Amending kerosene or heating fuel with 20% gasoline, or as much as 30% in colder weather, is necessary to get efficient use from a drip torch. Also, fire-lines can be maintained by dragging the burning duff ahead with an ordinary garden rake. If the crew splits into two parts, such as circling the site from opposite directions, it is useful to have two-way radios available. Alternatively, two cell phones will be entirely satisfactory. Fire rakes, consisting of triangular metal teeth (not shown), usually are the least used tool during a fire, but are handy for raking firebreaks to bare soil if needed. Raked firebreaks are completely useless in the face of a raging headfire; thus, fire rakes are most useful in preparing a fire-line for a backfire before the burn.

Of all the tools, the backpack firefighting water pump requires the most maintenance. Do not expose the pumps to freezing, and when storing the pumps, make sure you pump out all the water and store the metal tanks upside down. Occasionally the pump slide requires oil, and a residue buildup on the sliding pumping cylinder can be cleaned with steel wool. Leaking valves requires disassembling the pump to correct the problem, which usually is the result of a faulty gasket. Spare parts are available from the supplier.

Supplies for all types of firefighting equipment can be obtained from Forestry Suppliers, Inc. (P.O. Box 8397, Jackson, MS 39284-8397; phone 800-647-5368; website www.forestry-suppliers.com) or Ben Meadows Company (P.O. Box 5277, Janesville, WI 53547, 401 South Wright Rd., Janesville, WI 53546; phone 800-241-6401, fax 800-628-2068; website www.benmeadows.com).

6.4.4 Required Clothing

Two basic rules of dress are (a) do not wear garments of artificial cloth, as they tend to either melt or ignite, and (b) wear adequate cover to prevent skin burns from radiant heat. Specifically, wear a hat, long-sleeved cotton or wool shirt, full-length cotton trousers (denim is good), gloves, and work or hiking boots. Glasses should be worn, but the full-eye protection variety of glasses is often not satisfactory, as they tend to fog over with trapped perspiration. A bandana for facial cover is optional. You will find that a water canteen is a welcome item. The added protection of fire-resistant outerwear is available from various forestry suppliers; however, this can be costly for occasional use.

6.4.5 Burn Strategies

Proper planning and favorable conditions are essential to achieve a safe prairie burn. Raging prairie headfires, that is, a grass fire burning with the wind, is a dangerous force generating flames between 5 and 30 ft high. Several facts about fire control should be kept constantly in mind. (a) Always start the fire downwind to create a firebreak against a headfire. (b) Priority planning should be given to securing buildings. (c) Take note of firebreaks, such as roads, wide trails (trails and even roads are useless in the face of a headfire), and streams. (d) Wind speed and relative humidity are extremely important. (e) Uphill burns are especially energetic and thus require extra caution. (f) Oak-hickory forests afford a partially secure firebreak, except on uphill slopes; that is, leaf fires can be swatted out as they burn about 6 in. in height. (g) Take into account utility lines for sensitivity to heat and possibility of igniting utility poles. (h) Designate an experienced person as a fire boss with authority over a burn. (i) Do not assume that a field with corn stubble is a safe firebreak because corn has been known to produce record amounts of dry matter in a season.

To be specific, there is more to a burn than the summary outlined above. Planning must begin months in advance. Prior to a burn, an unfamiliar site should be walked to identify potential problems and the location of firebreaks or lack thereof. A map is often handy. Mowed firebreaks should be completed well in advance. Mowing a firebreak 6–10 ft wide can make your burning task much easier, but a mowed firebreak is far from being totally secure. A more secure firebreak consists of two mowed swathes in parallel leaving an unmowed area between; then, sometime before the burn the unmowed swath can be carefully burned. One can mow in the fall before a planned spring burn or constantly mow the break on a regular basis. A scrupulously cleaned mower should be used (to avoid introduction of unwanted exotic seeds). Excess clippings should be raked about 10 ft into the prairie side of the break. If required, applications for burn permits from state and local agencies should be submitted well in advance. For example, Illinois requires a permit from IL EPA, and usually burns within city limits require a permit from the fire department. Although landowners in remote areas regularly burn without a permit, it is prudent to not run afoul of the authorities, especially within city limits. About 1 week before the burn, neighbors should be notified about your plans, so that the fire department is not called unnecessarily: they are required to respond to any call to 911. Then, the day before the planned burn, a detailed weather report for the next day should be evaluated for suitability. Updated weather can be obtained from the National Weather Service either by use of a weather radio or through the Internet. On the day of the burn you must call the non-emergency number of the fire department and explain your plans. However, expect that passing motorists may call the fire department on occasion, which is impossible to predict. The fire department often is required to check it out even though prior notification was received. As part of your prior planning, make sure your fire crew is in good health, especially with no respiration or heart conditions. A good portion of your crew should be experienced. If not, you should start your training with a small plot or volunteer to help an expert crew.

Now you are prepared to work, assuming that the weather conditions are satisfactory. The relative humidity should be in the range of 25–60%; however, above 50% a minimum of 10 mph wind is needed. Relative humidity often can be judged by air clarity and wind direction. Wind out of the northwest is a harbinger of low relative humidity, and winds out of the south or east usually bring in high-moisture air. Generally, an increase in air temperature by 20°F will lower the relative humidity by half. Thus, the hottest time of the day at about 3–5 p.m. usually will have the lowest relative humidity. Also for this reason, it is not advisable to burn above a temperature of 80°F. To avoid low relative humidity, plan your burn either for morning (keeping in mind that the day will develop progressively lower humidity) or early evening/dusk when the temperature begins to fall. On the other hand, at temperatures below 32°F it is difficult to maintain a burn, especially with light fuel. Wind plays a definite role during cold periods. I have successfully managed a burn in the face of light snow flurries being driven by a stiff wind out of the north. In that particular instance, the surrounding area was secure by plowed fields. Under normal conditions the wind speed should range between 5 and 15 mph. Below 5 mph conditions may be satisfactory in a secure area, as inconsistent wind direction will be a problem. Wind speeds above 20 mph invite disaster. It is prudent to have on hand a weather radio in the field.

If the weather conditions are acceptable, your next order of business is to check the wind direction in several places by throwing up a handful of grass and noting the result. With this in mind and all firebreaks in order, you will want to create a secure burned-out perimeter of at least 200 ft around any susceptible buildings in the area by backfiring (a fire moving against the wind). See below for backfiring techniques. Depending on the location of the buildings, you may want to put this backfire out before moving on. If extra help is available, post a firefighter at the building site. Alternatively, firebreaks around buildings can be completed before the main burn.

On a site with secure perimeters you are now ready to backfire at a downwind position. Relatively secure perimeters are defined as roads, water bodies such as streams, or a previously mowed or burned firebreak. Deciduous forest edges are not secure, but they can serve as a partial firebreak with a trained crew on hand; that is, leaf fires are manageable on level ground or downhill slopes. However, make sure you account for insecure spots upwind, in which case a small detail can be assigned at these positions. With all in readiness, a two-drip-torch strategy is best where the crew breaks into two groups working in opposite directions, both burning against the wind. Unlike the intensity of headfires, a backfire generally burns 1 ft wide × 1 ft high. Beside the drip-torch, each team should have a couple of persons manned with either swatters, shovels, or fire-brooms, including a person with a backpack water pump to make certain the fire is controlled. Water should only be used when necessary. Rely mainly on swatters to put out the fire. Obviously, 5 gallons of water will not far without replenishment. In this manner, the entire prairie can be encircled with the upwind portion being saved for last. Completing the whole site with only a back-burn is another option. Figure 6.2 shows a prairie fire started downwind, then encircled in opposite directions by two teams, including one person on each team carrying a drip-torch. Persons carrying fire-flappers and backpack water tanks followed the drip-torch carriers. Cell phones or two-way radios are recommended to coordinate the final enclosure, and fires at right angles to the wind will have burned a safe zone (at least 200 ft) before the intense head-fire races ahead with a wall of flame 5–30 ft high in a strip 5–15 ft wide. Do not ever get in the path of a headfire; they are virtually unstoppable.

Figure 6.3 illustrates a fire crew of ten persons, which is adequate for a 60-acre prairie burn surrounded by oak forests. Typically, two crews of five to six persons start against the wind and back-burn in opposite directions until they meet after completing the perimeter. At this time, a head-fire fueled by the wind is safely accomplished.

Fig. 6.2 Prairie fire started by the encircling method

Fig. 6.3 The Jubilee Park "Prairie Dawgs" relax after a 60-acre burn

Some comment about burning firebreaks is appropriate. Burning firebreaks before the main burn is a very tricky and difficult job, but they can be accomplished with a trained crew. Personally, I prefer to have a mowed path at least 6–10 ft wide downwind prior to starting the fire. If the firebreak is needed to secure a smaller area to be burned in a larger expanse of grass, there may not be another choice. If the firebreak is extensive, a day before the main burn can be scheduled. When relative humidity is rising around dusk, the chore of burning firebreaks is simplified. Early mornings are subject to decreasing relative humidity and increased wind speed as the day progresses, thus the work should end when the fire becomes difficult to control. Adhere to the same rule of starting downwind when burning firebreaks. If the wind is out of the west, locate your crew with swatters and water backpacks on the east. Start at one end and fire about 10–20 ft of grass toward the north (or south). Allow the flame to burn against the wind toward the west until it is about 10–20 ft wide. After this, the flame must be immediately swatted out on both sides. Then move a line of fire an additional 10–40 ft further on. If the backfire is moving too slow, an additional line of fire might be set parallel to the first about 5 ft upwind. Continue with this process until the downwind portion is finished. Next, the tricky business of firing a firebreak at right angles to the wind is started at one end of the downwind firebreak. If the wind is out of the west, you will be moving in a straight line from east to west by starting a fire about 5 ft wide at one end of the eastern firebreak. Permit this fire to back into the wind by stationing the crew on both sides of the back-burning fire, swatting it out when it is 5–20 ft wide. The torch-bearing person should move toward the wind only so fast as to allow the swatters to catch up. To prevent a headfire, do not permit a bulge to develop. Repeating this procedure on the opposite end of the eastern firebreak will complete the U-shaped firebreak. Do not enclose the U to create a headfire, as this fire will overwhelm a mere 20-ft-wide firebreak downwind. After the firebreaks are completed, then you go back to the downwind position to start backfires. After the backfire widens sufficiently to about 100–200 ft, two teams should then encircle the inside of the firebreak, moving toward the wind. Make sure the two teams keep abreast. If the wind shifts, the downwind team should move ahead of the other team to avoid smoke and a headfire. When the firebreak has been widened on both sides by this procedure, the circle can be enclosed to allow the headfire to roar ahead to complete the job.

With a lack of help to split into two encircling teams, two other alternatives can be employed. First, completing the job with only a backfire is a slow but relatively safe alternative. In the second alternative, the backfire is allowed to burn about 100–200 ft wide, then the torch person burns a strip 20 ft ahead of the backfire, causing a relatively short burst of intense headfire. Depending on the comfort level, the strips can be increased up to 150 ft, but make sure the firebreaks on the two sides are safely holding.

Hillsides present a special problem. The most intense fire I ever witnessed was a rapidly torched bottom of a hill. The fire literally roared uphill generating huge flames, many whirlwinds, and much billowing smoke. Although I was standing 100 yards from the line of fire, I felt compelled to stand behind the trunk of a large burr oak. The fire burned out as rapidly as it started. This burn was accomplished relatively safely because the hilltop dropped off a 20-ft cliff to the mainline of a double-tracked Santa Fe Railroad. If a relatively steep hillside is involved near a fire-line, you must burn the firebreak on the very crest of the hill, regardless of the direction of the wind. Fire will usually move safely downhill, acting as if it were a back-fire. In the East, an area to be burned often has several hills with different orientations. Figure 6.4 shows the aftermath of a hilly burn in Pennsylvania. In such situations, it is strongly advised to separate the parcel into different sections with mowed firebreaks as shown.

Fig. 6.4 Aftermath of a burn in an Eastern hilly terrain: note the mowed firebreaks surrounding sections of differing hillside orientations (*arrows* show directions of burn)

A few additional potential hazards should be mentioned. Standing marsh water with grass interspersed is not a reliable firebreak. Fire can race through cattails, cord-grass, or reed canary grass with ease. In bogs beware of dried peaty soil as peat fires are impossibly difficult. For example, a disastrous peat fire occurred at the Green River in Illinois as a result of a prescribed burn. Attention to utility poles was mentioned above. Old utility poles displaying gray-colored wood are especially susceptible. Likewise, old gray fenceposts easily ignite. Newer treated posts are far more resistant to burning. As overhead communication lines improve technologically, the chance of heat damage increases. It is appropriate to avoid these utilities or back-burn under the lines.

As mentioned before, oak/hickory wooded edges can act as a partial firebreak. Burning leaves are managed by swatting out the relatively low flames; however, note that uphill leaf fires can be surprisingly rapid and treacherous. Large oak and hickory trees are largely resistant to fire, but lower branches are killed. Smaller oak/hickory trees are killed above ground but continue to repeatedly sprout new growth the next season. If you want to ensure the health of larger savanna trees or prevent killing the aboveground portion of smaller trees, you must mow a firebreak around them and rake away all dry matter. If possible, it is preferable to back-burn a reasonably large radius around them. Headfires will cause surprising damage at a fairly long distance. Early succession trees, such as chokecherry, box elder, and dogwood, are easy to kill, and repeated burning leads to their demise. Similarly, brushy growth will disappear with repeated burns, especially if the brush is growing in the midst of prairie. Highly desirable woody prairie shrubs, such as *Ceanothus americanus* (New Jersey tea) and *Amorpha canescens* (lead plant) are adapted to prairie fires; thus, they will resprout anew from the root. However, it is the opinion of this author that repeated annual burns inhibit these two species. Thickets rarely burn because there is little flammable litter on the ground to maintain a flame. If pine trees or cedars are growing on the property, be apprised of the possibility of crown fires, which would be a disaster in dense stands of conifers. Crown fires in shingle oaks occasionally have surprised our crews. Shingle oaks do not lose their dead leaves until late spring. Such fires burn out quickly, and the trees do not appear to be notably damaged. Burns are an excellent method to eventually eliminate poison ivy, but keep persons sensitive to poison ivy away from the smoke.

You may choose to permit the fire to burn the leaf litter through the forest. Historically Midwestern prairie fires swept through oak/hickory forests, creating open savannas. Fire changes the character of forests by killing early succession trees, brush, and fire-sensitive trees such as maple. Additionally, the lower branches of fire-resistant trees are killed, thus opening the canopy to promote the sprouting of acorns and the growth of partial shade prairie plants. The vistas through the forest are increased considerably, giving the impression of an ancient forest. When the burn area is moderately secure, such as areas surrounded by streams or roads, we burn the entire ecosystem, including the forest. Such burns causes fires in dead or punk wood that can last for a week. Chimney trees are especially spectacular when the hollow trunk of a dead tree can act as a fire-conducting chimney. Burning trees are not a problem if surrounded by a burned-out area or otherwise secured by firebreaks. Be sure chimney trees do not collapse over a firebreak.

An escaped wildfire is the nightmare that none wants to experience. Under these circumstances, you will want a cell phone to call the fire department for help. Hastily, but not carelessly, you will want to assess the situation for natural firebreaks and the possibility of heading off a headfire with a back-fire. This strategy require advancing quite a large distance around the headfire to light an intercepting back-fire. By all means keep out of the direct path of a headfire. Alternatively, two teams can work on the flanks to narrow the advance of the flame.

Further information on managing prairie fires can be obtained by reading Pauly [29].

6.4.6 Murphy's Law: What Went Wrong

After experiencing thousands of acres of prairie burns ranging from less than an acre to 200+ acres, some personal accounts of mistaken judgment can be instructive.

My first burn was a forceful way of learning respect for grass fires. Being only 100 ft square, the plot of big bluestem and Indian grass at the edge of my lawn appeared rather innocuous, and thus no tools were immediately handy. Lit by one match, the plot burst into a roaring headfire that sent me scrambling for a shovel and a hose. By the time I returned, the fire fortunately had rapidly burned itself out at the edge of a brushy area.

In another incident at a 10-acre prairie, a back-fire was started by our crew on the downside of a west wind on an eastern mowed firebreak. A sudden storm swept in from the south/southwest. The wind increased notably from the south, and the back-fire suddenly changed directions. A rapid headfire swept through a narrow section of oak trees adjacent to a road on the north. Burning leaves jumped the road, igniting the roadside on the opposite side. If the field had not been plowed, there would have been an uncontrolled fire. This situation may have been avoided by the use of an on-site weather radio. One should not start a grass fire with thunder or thunderclouds on the horizon, especially in the west or southwest. An additional lesson was learned: roads are not always reliable firebreaks.

As a helper on a burn managed by the U.S. Army Corps of Engineers, several near-disasters occurred. After tiring the crew (some untrained) by trying to control a 50- to 100-acre fire from racing uphill into an oak forest and a suburb beyond, we encountered a burning utility pole supporting a high-voltage line. Because the fire had reached about 20 ft high on the pole, it was impossibly high for swatting tools. Fortunately, a mostly empty backpack water pump reached the flames using the one-hole nozzle option. In another near-disaster, a Corps truck was parked on a mowed area near the tallgrass fire. Despite the apparent safety of a mowed lawn, the grass was dry, causing the fire to sweep under the truck. Because the keys were not left in the truck, the vehicle could not be moved. Extreme crew effort of swatting a fire under the vehicle saved the day. Mowed lawns are not always safe firebreaks, especially those composed of warm season grasses, such as buffalo grass.

On another burn, I had an extremely exciting time swatting out a buffalo grass lawn fire that sent tongues of fire spreading rapidly in many directions.

Occasionally, prairie fires penetrate forests too far. Under these circumstances, it can be difficult to put out smoldering wood fires. It is best to prevent the fire from advancing very far into the forest. Although the forest edge can be used as firebreak, penetration can be a headache. One of our burn sites was a 60- to 80-acre prairie completely surrounded by forest. After we completed a burn on this site, we hiked out a half-mile to enjoy refreshments of cold beer and sandwiches. To our dismay, smoke on the horizon indicated that the fire had been re-ignited by smoldering wood at the forest edge. Although the fire crew was exhausted, a small group had to hike back in to swat out leaf fires on forest hillsides. In another year the fire was permitted to penetrate excessively into the forest because the drip-torch person had proceeded too far ahead of the fire crew. The crew had to work into the evening hours securing the area from burning deadwood. In securing burning wood, it is often easier to move the burning wood into a burned-out area or create a burned-out area around the deadwood. Because prairie grass dries more rapidly than forest leaves, such burns can be simplified by scheduling after a previous day's light rain.

Fires in metropolitan areas are especially challenging. We managed a prairie restoration site at a USDA lab adjacent to an interstate highway and close to downtown Peoria, Illinois. First of all, fire permits had to be obtained from Illinois EPA and the Peoria Fire Department. Before the burn, a firebreak was made along the property line of the Peoria baseball stadium. On the day of the burn, notification was given to the fire department, the lab safety officer, and the lab director. A fire was ignited first along this property line only when the wind was away from the stadium and the interstate highway. After all these precautions, a fire-created whirlwind sent a burning prairie dock leaf sailing into a stand of Hungarian brome on the stadium side of the fence. Illustrating the utility of ample help, four firefighters immediately jumped the fence and pounced on the small patch of fire to save the day. In subsequent burns, two persons with backpack firefighting pumps were stationed on the other side of the fence even though it seemed to be an unnecessary practice.

On occasion inadequately trained members of the fire crew have severely damaged fire equipment. Dropping equipment in tallgrass to help another usually results in burned equipment. Backpack water pumps are susceptible to melted shoulder straps and damaged vinyl bags, and on one occasion a drip-torch was recovered just in time to avoid disaster. If you drop equipment, make sure it is located in a burned-out area. Also, the careless resting of rubber flappers in fire during use can cause a damaging burn. Constant use for the purpose of flapping out a fire rarely results in damage. Do not allow the flapper to linger for long directly in flames, which often happens with inexperienced crews.

6.4.7 Alternative to the Use of Fire

If fire cannot be used because of safety concerns, such as location in an urban setting, a satisfactory alternative would be fall mowing after seed set of late-blooming plants.

This time can range from mid-November to late March before early plants emerge. This procedure will inhibit the dominance of woody trees and brush and improve the appearance of early season forbs, which are often obscured by dead vegetation of the previous growing season. Because mowers are notorious for the spread of seeds of alien plants, make sure the mower is scrupulously clean. Persistent problems with woody vegetation can be solved with the procedure of painting individual cut stubs with 18% or greater glyphosate (isopropylamine salt) (see Sect. 6.5).

6.5 Control of Alien Plants

Control of Eurasian invaders of native ecosystems is an important component of restoration efforts. Because some of our North American natives have become aggressive aliens in Europe and Asia, ecosystem degradation is universal. Invasion of exotic alien plants are an increasingly serious problem to land managers, if not a nearly hopeless situation. Alien plant invasion in the U.S. is extending even into remote areas from Hawaii to the East Coast. Currently, this author lives in Pennsylvania where two and a half centuries of agricultural and horticultural introductions have transformed the area into a haven for alien plants. Only long-lived trees persist as dominant natives. Central Pennsylvania is on the edge of incursion by mile-a-minute vine, *Polygonum perfoliatum.* A reading of Tallamy [9], pages 78–80, will convince anyone that an invasion by Martians would be preferable. Equally serious are imported animals, insects, and plant diseases. We recognize the untimely fate of the American chestnut and American elm. Now alien diseases or insects threaten hemlock, dogwood, ash, and oak.

Largely at fault was the unregulated importation of exotic aliens by the horticultural industry, which continues today. Nursery owners will only change if the public starts demanding only native plants. Accidental introductions from imported agricultural commodities are another source. Also, the Natural Resources Conservation Service of the USDA was responsible for importing several nasty aggressors to "stop soil erosion." The Forest Service (also a division of USDA) is particularly alarmed by the spread of aggressive alien plants [30]. In an ironic turn of events, the Forest Service is trying to eliminate aggressive African buffelgrass, *Pennisetum ciliare*, from southwestern states, while another division of USDA is working to develop a cold-hardy strain for use as forage! It is difficult to refrain from adding editorial remarks about the USDA bureaucracy. In 2000 Cornell University researchers estimated that invasive weeds cost the public $138 billion annually, and the costs continue to grow. The Clinton Administration recognized alien species as serious to the extent that the National Invasive Species Council was formed in 1999. This Council created a National Invasive Species Management Plan [30].

Not all alien plants are problematic, but many alien plants become aggressive partly because their natural enemies are absent. Some of the aggressive aliens take advantage of disturbed soil to "squat" on the land but are inhibited by well-established native perennials. For example, this author has observed the aggressive alien *Bromus inermis*, Hungarian brome, growing on a roadside immediately adjacent to a pristine Wisconsin Nature Preserve. Neither plant community invaded the other, as if a line were drawn in the sand. Perennial plants, especially, hold on to their territorial rights. If permitted to lie fallow for a few years, Eurasian perennials invade abandoned farm fields. For this reason, intensively cultivated soil treated annually with herbicide is preferred for restoration efforts, provided that the applied herbicides are not persistent. Fields overcome with perennial weeds should be thoroughly denuded with a stronger than normal dose of the nonpersistent herbicide, isopropylamine salt of glyphosate, sold variously as Round-up (16 and 60%) or Ultra-Kill (25–41%). A second or third round of herbicide may be necessary to guarantee that patches of alien plants are not missed. Working the soil by plowing, harrowing, or roto-tilling before a kill by herbicide is totally unacceptable. For example, working the soil will break up the rhizomes of *Bromus inermis* to subsequently give a lush growth from the fragmented rhizomes. After making certain there is total kill of perennial weeds, one can make a single pass with either a roto-tiller or harrow to prepare the soil for native seeds. Sowing directly on herbicide-killed thatch has the advantage of preventing buried weed seeds from surfacing. After a prairie restoration is completed, *do not use a mower unless it is dedicated only to the restoration plot.* Mowers are efficient carriers of weed seeds; a total decontamination of the mower from debris is necessary.

Some of the most notorious alien or non-prairie invaders are described below in rough order of their aggressive nature in Midwestern and Eastern prairies. There are an increasing number of aggressive alien plants; thus, the plants listed here give an overview of the ones most commonly encountered in prairies. Other non-native plants may be locally problematic. The control measures described below cite many of the recommendations of the Illinois Nature Preserves Commission's guideline on alien weed control [31]; also see M.K. Solecki in Packard and Mutel [10]. These references give additional information, particularly herbicide alternatives. Alien plants should be positively identified before planning their elimination. No matter the effort necessary, this author is a strong proponent of aggressive management of alien weeds. Otherwise, you will quickly realize that your restored ecosystem has become degraded. To a large extent a mature native ecosystem will exclude unwanted aliens, but developing this ecology from scratch can be a challenge.

Several herbicide concentrations are mentioned in the sections below, requiring calculations for the formulations. For example, the manufacturers of glyphosate (isopropylamine salt; see footnote 1) usually recommend 6 fluid ounces (oz.) of 18% glyphosate to be added to 1 gallon of water. One gallon of water amounts to 128 fluid oz., so the mixture would be $128+6$ oz. $=134$ oz. The 6 oz. of 18% glyphosate would contain $6\times0.18=1.08$ oz. of pure glyphosate. Thus, the final concentration of glyphosate added to 1 gallon of water would be 1.08 oz. glyphosate per 134 oz. water + herbicide = 0.0081 oz. glyphosate/oz. of mixture. Converting to percent: $0.0081\times100=0.81\%$ glyphosate. Ordinarily, the manufacturer's recommendations can be followed, except when higher concentrations are required for trees and brush.

1. Trees and brush: In early days of colonization of the Midwest, large expanses of tallgrass prairie were completely devoid of trees. In the absence of a source of tree seeds, invasion by trees was rare. In present times, it is a vexing problem. Early succession trees, such as wild cherry, mulberry, maples, elm, tree of heaven, and locust are targets for clearing. Large fire-resistant trees (late succession trees), such as certain oaks and shag-bark hickory, are natural components of savannas, and as such should be permitted to remain, unless they threaten to overwhelm the prairie. In this latter category, selective clearing can lead to a desirable savanna ecosystem. Bur and white oak as well as shagbark hickory are relatively resistant to fire, but others, such as red oak, are more sensitive to fire. Regular burns give good control of tree growth, especially those categorized as early succession. As a general rule, prairie fires are a good method of controlling woody vegetation. Small native trees are often found in the prairie, such as American plum (*Prunus americana*), haw/thorn/hawthorn (*Crataegus* sp.), dogwood (*Cornus* sp.), and hazel-nut (*Corylus americana*). Although the aboveground parts of these smaller trees are also killed by fire, they usually regenerate from the root. Repetitive burning eventually is harmful to their survival. If thickets of brush and early succession trees infest the site, it is very difficult to generate a killing fire, because ground litter is deficient. Thus, other methods must be employed. Axe girdling can kill large early succession and/or alien trees. Applying full-strength glyphosate (18% or greater) to the girdle is even more effective. Depending on the bark thickness, some trees can be killed by girdling the tree using a propane torch to kill the cambium layer. Herbicide treatment is another alternative. Garlon herbicide is often recommended for trees. Because glyphosate is a relatively nontoxic, nonpersistent herbicide, I prefer its use. A couple of downward-slanting holes can be drilled into the trunk of large trees, followed by pouring undiluted 18% glyphosate into the holes with the aid of a funnel. Saplings and brush can be cut with lopping shears or a brush cutter, followed by painting 18% or greater glyphosate on the cut stump using a paintbrush or a sponge applicator. Doug Franks (personal communication) adds Metamucil to glyphosate as a thickener. A two-person team composed of a cutter and a herbicide applicator is preferable. Addition of food coloring can aid detection of stumps already treated. Although undiluted glyphosate is effective, the Illinois Nature Preserves Commission advises that other concentrations are satisfactory [31]. The following concentrations are recommended for various species, with the precaution that various concentrates of glyphosate are commercially available (usually 15–60%): autumn olive (*Elaeagnus umbellata*) 10–20%, black locust (*Robinia pseudoacacia*) 20%, buckthorn (*Rhamnus* sp.) 50%, bush honeysuckles (*Lonicera* sp.) 20%, multiflora rose (*Rosa multiflora*) 10–20%, quaking aspen (*Populus tremuloides*) 10–20%, Siberian elm (*Ulmus pumila*) 10–20%, and smooth sumac (*Rhus glabra*) 10–20%. In the East and to some extent in the Midwest, privet (*Ligustrum vulgare*) is an aggressive bush, which can be eliminated by cutting and painting the stump with 41% or greater glyphosate; smaller privet can be easily uprooted by hand. Especial attention should be given to eliminating extremely aggressive alien trees, such as tree of

heaven, *Ailanthus altissima*. Mulberry trees, *Morus* sp., are particularly vexing as berry-eating birds deposit seeds prolifically. Burning the prairie keeps the aboveground portion of the mulberry saplings at bay, but a huge taproot continues to develop. It is recommended to search out the "mother" trees for girdling and painting the girdle with 18% or greater glyphosate.

2. *Bromus inermis*, Hungarian or smooth brome: This cool season perennial grass is very opportunistic. Hungarian brome spreads rapidly, forming large colonies with a tight mat of rhizomes. This tight mat will make the introduction of other plants very difficult. New colonies of brome form seed-heads, which also aid its propagation. Because tilling the soil merely fragments the live rhizomes, this method is totally ineffective. It is very important to get a complete kill with glyphosate herbicide before restoration. The strength of the mix should be about twice stronger than recommended for general weed kill (about 1.5%). Herbicide treatment should be completed in the spring or fall when this cool season grass is growing most actively. More than one herbicide treatment may be necessary to kill the survivors. Often living rhizomes remain that reinfect the area. If possible, herbicide treatment should be completed twice, preferably in the spring and fall or vice versa. Interestingly, many prairie species, such as the tallgrasses, are late season plants. In mixed brome and prairie colonies, the late season prairie plants will usually survive herbicide treatment in the spring. As a general rule, a successful prairie restoration will exclude the return of Hungarian brome, especially with a fair density of prairie grass. Hungarian brome seems to prefer partial shade; thus, competition with full-sun prairie is probably not the preferred habitat for brome. Because Hungarian brome is a cool season grass, some experts believe that late prairie burns about the first part to the middle of April (in northern states, mid-April to early May) inhibit brome in favor of prairie, but this is not an immediate solution, and late burns also inhibit early season forbs. Successful restorations have been reported by direct seeding over brome, but this method may take up to 20 years with regular spring burns. Hungarian brome seems to be regionally aggressive, as it thrives less well in more southerly latitudes. Unexpectedly, I have seen *Gentiana alba* spot-colonize healthy stands of Hungarian brome.
3. *Melilotus officinalis*, white and yellow sweet clover: White sweet clover was formerly known as *Melilotus alba*, but now it is categorized as a variety of *M. officinalis*. Inasmuch as white sweet clover is a larger plant than yellow sweet clover, it is more of a competitor in restorations in my opinion. Initial infections can occur from contaminated mowers, or as a minor component of commercial seed, such as red clover seed. Sweet clover is a biennial, and consequently infestations sometimes occur in 2-year cycles. In the first year of growth the plant is inconspicuously growing a large taproot, which it uses the second year to grow rapidly, flower, and set seeds. Sweet clover should be pulled during the early stages of flowering when it is easy to identify but before seed set. After seed set, the pulled plants must be removed from the site. In any case, make sure the plants are removed before the seeds mature and drop. The Illinois Nature Preserves Commission (IL NPC) [31] recommends pulling the plant in the fall

after its first year of growth; however, the plant is not easy to spot from a distance, and wet conditions are necessary to pull the swollen taproot. During flowering the taproot shrinks somewhat and the stem becomes more fibrous, making the job of uprooting easier. Sweet clover is a particularly troublesome weed as thick stands can be devastating to a healthy prairie. According to the IL NPC, sweet clover seeds can lie dormant for 30 years, making a continuing effort mandatory. Prairie burns can trigger the germination of sweet clover seeds by either seed scarification or increased light exposure, which becomes evident when many plants bloom the second growing season after a burn.

As recommended by the IL NPC [31], two alternatives can be used to control dense stands of sweet clover. One method requires cutting the stems close to the ground with a hand-held weed-cutter after the leaves on the lower stems have died, either before flowering or in the early stages of flowering. If cut close enough to the ground, sweet clover generally does not recover. According to Stephen Packard of The Nature Conservancy, "what is critical is cutting lower than the lowest senesced (dead) leaf." A second method requires a 2-year cycle of prescribed burns. The first-year burn should be completed in April, followed by a second-year burn in May. The first burn stimulates seed germination, and the second late burn kills emerging shoots. This procedure should be done twice, separated by 2 years without burning. The burn schedule has two disadvantages: (a) a late burn may be patchy, and (b) early emerging prairie forbs would suffer a setback.

Extremely dense stands of sweet clover may be too difficult to weed by other methods, in which case herbicide may be considered. However, herbicide should not be used in high-quality prairie; thus, the alternatives described above may be much more suitable. Publications of the IL NPC [31] recommend a formulation of 2,4-dichloro-phenoxyacetic acid (2,4-D) and/or Dicamba (3,6-dichloro-2-methoxy-benzoic acid) herbicide with an amine to prevent herbicide drift, such as Mecamine [*N*-2,3,3-tetramethylbicyclo(2.2.1)heptan-2-amine], also known as mecamylamine. The formulation should be a 1% solution of Mecamine and the herbicides, 2,4-D plus Dicamba. Because all broadleaf plants are susceptible, spraying should be restricted only to spot-spraying of sweet clover. Following a fall burn, a spot-spray application should be made in the spring on early emerging sweet clover before native prairie plants appear; this will kill surviving second-year plants as well as emerging seedlings. Spraying should wet the sweet clover leaves or seedlings but not drip off onto nontarget plants. This method would be particularly suitable for stands of prairie grass resistant to 2,4-D.

4. *Pastinaca sativa*, wild parsnip: Wild parsnip, an edible root crop gone wild, is a nominal biennial. Because its colonies are occasional in the East, it is a curiosity how parsnip has spread so efficiently over the Midwest. Perhaps Midwest soil is ideal for parsnip growth, but one wonders if parsnip was planted widely in earlier times to provide hogs with winter food. Roadside mowing is an effective way of spreading the seed, and one usually finds the healthiest colonies along freeway right-of-ways. The largest density of parsnips in prairies occurs in the most recently disturbed areas and partial shade near trees. As parsnip seeds germinate

over a period of about 5 years and parsnip competes somewhat with prairie plants, the best option is pulling the plants up by the root. The best time to pull parsnip is when the plant is in full flower before the head goes to seed. After the seed-heads develop, the plant must be removed from the site. Flowering shrinks the root sufficiently to permit its removal without too much pulling effort, except for the larger roots, in which case a tool used to root out dandelions will suffice. Additionally, it is very difficult to locate the plant before flowering. The area must be checked for several weeks because not all the plants flower at once. It is less effective to cut off the flowering top, as a new flower will develop from the root, albeit a much smaller flower. If cutting is chosen, one must return to cut the small blooms. Tedious uprooting will bear rewards after 5 years, after which only occasional parsnips will need to be pulled. Parsnip should not be pulled unless one is fully protected from direct contact with the plant. Over the more than 100 years since parsnip has gone wild, this plant has developed a high concentration of toxic chemicals called furanocoumarins. Contact with the skin and subsequent exposure to ultraviolet light from the sun causes severe painful blistering of the skin. A long-sleeved shirt, trousers (not shorts), and a substantial pair of gloves are required. Rubber gloves used to handle chemicals are the best choice. Because parsnip is usually found occasionally throughout a restoration project, the use of herbicide risks damage to wanted species. It is far less risky and efficient to simply pull the plant when in flower.

5. *Dipsacus fullonum/sylvestris*, *D. laciniatus*, Fuller's/common teasel, and cut-leaved teasel: Teasel is a nominal biennial; that is, it exists as a basal rosette for 1 year or more, after which it flowers and dies. Teasel has invaded all temperate areas of the US, originating from its introduction from Europe in the New England area. Each teasel plant can produce 2,000 seeds. Thus, teasel invades new areas along roadsides, undoubtedly from distribution of seed by roadside mowers. Teasel also is found growing along streams from seed carried downstream. This plant is particularly difficult to eradicate. To prevent damage to established prairies, uprooting with a mattock on an annual basis is recommended; insufficient root removal can result in resprouting. Do not allow the teasel to develop a seed-head. Because cut flower stems tend to develop seeds, they should be removed from the site. Simply cutting off the flower stem often is not satisfactory, as smaller new flower(s) will often develop from the root. To prevent seed accumulation in the soil, the best solution is immediate eradication of solitary plants as soon as they appear. Hopelessly thick stands can be eradicated by a 1.5% solution of glyphosate herbicide before the plants flower. The amine salt of 2,4-D is also a satisfactory herbicide. Herbicide application is recommended in the early spring or late fall when basal rosettes of teasel are still green, while native plants are dormant. Because dormant teasel seeds continue to germinate for a number of years, the infected area must be monitored regularly.
6. *Securigera/Coronilla varia*, crown vetch: Crown vetch has been extensively used in roadside plantings for its putative value in holding soil. Although it is proving to be a bane of the earth, the seed can still be purchased from seed companies. Crown vetch is extremely difficult to eliminate. Large colonies force out other vegetation.

It has been reported that several years of either burns or mowing in late spring will control crown vetch [31]; however, this has not been in agreement with my experience. Multiple herbicide treatments are required for many years. The optimum times are in the early spring or early summer before flowering and then again in the fall. It is evident that several years of herbicide treatment may be necessary to overcome the germination of dormant seeds. It has been my experience that the broadleaf herbicide Ortho Weed-Be-Gone Max (2 oz. per gallon water) is effective as a regular treatment. The IL Nature Preserves Commission [31] recommends a choice of three herbicides, either 1–2% glyphosate, dimethylamine salt of 2,4-D used according to the label, or a 1% solution of mecamine salt of 2,4-D plus Dicamba. Because the root is rather large, uprooting the plant does not seem practical. In my opinion, an area contaminated with crown vetch is one of the most vexing challenges in restoration. Fortunately, the colonies do not expand especially rapidly. As indicated by its spread along highway right-of-ways, mowing definitely spreads the seeds to new sites. One alternative is overseeding with tallgrass after glyphosate treatment followed by annual spraying with a 2,4-D formulation. This method would permit aggressive competition by 2,4-D-resistant tallgrass.

7. *Lonicera japonica*, Japanese honeysuckle: Japanese honeysuckle is largely absent in many parts of the Midwest, but it is found nearly everywhere in the East. It is an aggressive vine that yields incredibly fragrant flowers in early summer, which persuades the unsuspecting gardener to cultivate the vine, only to regret it later. Japanese honeysuckle gains a hold in disturbed areas, such as along roads, and in brushy situations, from which it spreads outwardly by vines even into pristine ecosystems. Dense mature forests tend to exclude this honeysuckle, except in instances where the vine has managed to entwine and climb a tree. New roots will grow from vines, establishing new plants in the process. The spread into established ecosystems is slow but certain. Regular spring burns are a useful method of temporary control, but burning only destroys the aboveground vines. Regrowth from the root is inevitable. Likewise, regular mowing (with a clean mower) or weed whip can only temporarily prevent a choking tangle of vines. A semipermanent solution requires more drastic action. When Japanese honeysuckle is growing in areas containing native plants of some redeeming value, the vine can be uprooted, which provides an excellent relief until the vine again invades from elsewhere. In instances where the plant has a substantial woody base, a mattock can be used to root it out. If the area is hopelessly overcome with matted vines composed of a honeysuckle monoculture, then a 1.5–2% solution of glyphosate herbicide mixed with 2,4-D or its mimic, such as Ortho Weed-Be-Gone Max, should be used according to container directions. This effective mix should be applied after other vegetation has become dormant in the autumn, before a hard freeze of about 25°F. This alien pest may be controlled, but it requires exceeding effort once it has overcome an area. Another aggressive honeysuckle, *Lonicera tatarica*, Tartarian honeysuckle, is a brushy plant 5–10 ft in height that can be eradicated by cutting and painting the stump with 18–41% glyphosate. This bushy plant is also susceptible to fire, but the root is not killed. Small plants 1–2 ft in height are easily uprooted.

8. *Centaurea* sp., knapweed: There are about 30 alien *Centaurea* species, and a few native species exist. The most widespread alien species is spotted knapweed, *Centaurea stoebe/maculosa*, including its subspecies *micranthos*. There are eight alien knapweeds that have a spotty distribution, but have not yet spread widely, namely *Acroptilon/Centaurea repens*, Russian knapweed; *Centaurea cyanus*, garden cornflower; *Centaurea diffusa*, diffuse knapweed; *Centaurea jacea*, brownray knapweed; *Centaurea nigra*, lesser knapweed; *Centaurea nigescens*, tyrol knapweed; *Centaurea scabiosa*, greater knapweed; and *Centaurea solstitalis*, yellow star thistle. The most aggressive of the knapweeds is *C. stoebe*, which produces about 1,000 seeds per plant. Spotted knapweed is reported to be either a biennial or perennial. It is my experience that it is a short-lived perennial. Spotted knapweed is a serious problem on well-drained dry-mesic to dry hillsides where is gains a competitive edge by its taproot and probable resistance to transpiration because of its pubescent leaves. In the presence of grass, spot-spraying of a broadleaf herbicide, such as Ortho Weed-be-gone Max, is only partially effective; that is, many deformed individuals manage to survive and flower at a later time. Spot-spraying with 1% glyphosate before flowering in April to May is more effective, with some collateral damage to natives. In the presence of mainly natives, hand-weeding the flowering plant takes several years because of seed survival. Flowering plants must be removed. The taproot yields only in moist soil when the plant is flowering, or one can use a dandelion digger. Larger plants require uprooting with the aid of a mattock. Weed-whipping the flowering plant is an alternative that prevents the spread of seeds, but the cut plant should be bagged and removed to prevent seed maturation on the cut plant. Similar to many other alien weeds, knapweed spreads into fields from roadsides; thus, early detection is a wise choice.
9. *Phalaris arundinacea*, reed canary grass: Reed canary grass has the distinction of being both a native and an alien that reportedly is extremely difficult or impossible to distinguish. In any case, this aggressive grass should be removed before it degrades the natural diversity. It prefers wet or wet-mesic situations, especially in soil of high nitrogen content. Reed canary grass also can be found on mesic soils; it will tolerate both full sun and partial shade. This grass spreads readily by rhizomes forming expanding colonies. Some control can be obtained by uprooting the plant, but it is too difficult for large colonies. The method of choice is an early spring application of glyphosate herbicide on dry land or Rodeo in wetlands. Early spring is chosen because reed canary grass is green before most native wetland species emerge.
10. *Lythrum salicaria*, purple loosestrife: Purple loosestrife probably had its start from gardens planted with nursery stock. Care should be taken to not confuse this plant with the smaller native, *Lythrum alatum* (smaller in height, leaves, and flower). *L. salicaria* is spreading from the East toward the Midwest, where it has already arrived in certain Midwest locales. Purple loosestrife hopelessly infects wetlands, where it can become a solid sea of purple forcing out native wetland species. Early detection is the key. Smaller numbers of plants can be pulled by hand before the plant goes to seed. A tool may be required for older plants. Because the pulled plants and soil may contain seeds, bag the plants for

burning after removal from the site. A wetland version of glyphosate, namely Rodeo, can be carefully used as a 1.5% solution [31]. Like glyphosate, Rodeo is a nonspecific herbicide capable of killing all nontarget and target plants. A specific purple loosestrife-feeding insect from Eurasia has been released in the U.S.; thus, a long-term solution could be at hand.

11. *Daucus carota*, Queen Anne's lace: Queen Anne's lace is another nominal biennial. Because isolated specimens usually are scattered throughout a site, it is preferable to uproot the flowering plant, which is easy to locate. The taproot is rather small compared to parsnip, so resistance to pulling is less. A few years pulling will give notable results. In my opinion *D. carota* is less of a problem than the other aggressive biennial, *Pastinaca sativa.* As prairie matures, Queen Anne's lace will decrease because of competition.
12. *Alliaria petiolata*, garlic mustard: The biennial garlic mustard is more of a problem in wooded areas where it can eliminate woodland wildflowers. It possibly acts as an effective competitor in woodlands because of its lush growth in the absence of natural enemies. In addition, garlic mustard roots excrete the plant toxins, benzylisothiocyanates and allylisothiocyanates, into the soil [32]. Because garlic mustard initially appears along trails and roadsides, the seeds undoubtedly are spread on animal hooves, such as those of deer and horses, and by humans and by roadside mowers. Plants also are initially observed along streams, indicating conveyance of the seed downstream. The plant does less well in full sun, but there it will take advantage of disturbed soil, such as woodchuck or badger holes. In prairies it can be found along plowed fields, in ditches, among annual weeds, and under isolated trees. It rarely competes well in established prairie ecosystems. Uprooting the flowering plant in late May to early June is the recommended procedure. If uprooting the plants before it flowers, make sure the top does not break off from the root, as it will regenerate. Uprooted garlic mustard has an uncanny ability to survive and set seed with a very minimum amount of water. During rainy seasons, uprooted plants will set seed lying on top of brush or on gravel roads! Therefore, the easy procedure of stripping off the flower head from the uprooted plant is the best option in preventing seed development. Bagging the plant is a much more tedious strategy. If a large quantity is involved, a compost heap will be an alternative, but expect survival of the plants on the top layers. *Obviously, one should eliminate garlic mustard before it expands over a larger area and before it becomes well established.* Garlic mustard seed can lie dormant for up to 9 years, making it imperative that weeding is an uninterrupted annual spring affair. Early spring prairie burns can reduce its growth, delay flowering, and kill some of the plants, but a recent study found that fires aid its proliferation. Observing the results of woodland burning, it is my opinion that fire exposes or scarifies the seed to promote its germination. Expect a greater than normal number of plants the year after a burn. If fires aid germination, fire may benefit by reducing the seed bank provided that the resurgent plants are subsequently eliminated. It has been reported that cutting the flowering plants at ground level results in 99% kill, and a higher cutting 10 cm above ground gives 71% kill and 98% reduction in seed

[33]. This author has found that cutting is less than satisfactory, possibly because the plants mature at different rates. Herbicide use in the spring is unsatisfactory in quality ecosystems, as the procedure has to be repeated annually for many years. In addition, herbicide causes more disturbances, which is favorable to garlic mustard proliferation. When most woodland plants are dormant in the late fall, glyphosate herbicide can be used as a 1–3% concentration to eliminate the still-green garlic mustard [33].

13. *Elymus/Agropyron repens*, quack-grass or knot-grass: This grass is truly aggressive in disturbed soil, forming mats of jointed rhizomes. Knot-grass is an apt name, but it is also known by a plethora of other common names, such as witch-grass, squitch, wickens, couch-grass, quitch-grass, twitch-grass, stroll, quichens, pond-grass, shelly-grass, dog-grass, blue-joint, shear-grass, slough-grass, Colorado blue-grass, or false wheat. If a stand of knot-grass is present on the property, expect this species to be a serious problem initially. Glyphosate herbicide treatment can eliminate part of the problem by killing the plant plus root rhizomes, but it will come back from the seed bank. However, several years of maturing prairie will eventually cause its demise. Prairie burns undoubtedly contribute to the decrease of this early season grass. A few stray plants will persist. Hand-weeding the seed-heads will hasten its disappearance. Considering the problem with other aggressive species, hand-weeding of this species often receives low priority.
14. *Saponaria officinalis*, bouncing bet: Widely distributed roadside populations in certain areas are a sure indication that roadside mowers have aided the dispersal of the spherical black seed. It has been my observation that contaminated mowers have infected the bicycle trailside of the Rock Island Trail Nature Preserve where the plant did not exist previously. Consistent pulling of the plant appears to control its spread. However, in the face of other aggressive alien plants, bouncing bet often receives a lower priority.
15. *Hesperis matronalis*, dame's rocket: Incredibly, dame's rocket is used in wildflower plantings along freeways and even in prairie restorations. Although the flower may be attractive, it is an alien with aggressive tendencies. This author has not had experience with well-established colonies of this plant. Because the bright pink, purple, or white flowering plant (0.6–1 m tall) is easy to spot, an occasional individual should be immediately weeded by hand to prevent spread of seed.
16. *Schedonorus/Festuca pratensis/elatior*, meadow fescue/English blue-grass: There are a number of alien fescues other than the species listed. There are also native *Festuca* species. Alien fescues can be a problem in fields where it was previously introduced. Total kill with glyphosate herbicide should precede restoration efforts. Because of the resistant matted rootstock of this fescue, individual strays can be removed only with a mattock. Another recommendation requires a 2- or 3-year succession of late spring burns. Glyphosate herbicide (1–2%) can be spot applied in early spring or late fall when fescue is green while natives are still dormant.
17. *Cirsium arvense*, Canada thistle: This perennial, imported from Europe, not Canada, forms colonies from its running rootstock. Initial infections probably

occur by wind-borne seed. Canada thistle is not commonly found in undisturbed prairies, possibly because it prefers disturbed soil. In prairies it seems to prefer wet-mesic situations. Three consecutive years of burns in late spring no earlier than between May and June are effective measures of control, but this procedure may be too harmful to early season natives. Frequent pulling with heavy gloves or cutting of plants in June, August, and September will starve the roots. Well-established colonies can be most easily disposed by wick application of a 1–2% solution of glyphosate herbicide in the spring. This author has noted a very significant decrease of Canada thistle as the prairie ecosystem matures.

18. *Poa pratensis*, Kentucky blue-grass: Because Kentucky blue-grass is alien, "Kentucky" is a misnomer. This grass is often a result of using a lawnmower in a prairie restoration. Seeds are easily transferred from the mower. The grass is most common where competition from prairie plants is less, such as along a path. However, Kentucky blue-grass can also exist as an understory plant. Because this grass is not very distracting, it can be ignored. Avoiding the use of contaminated lawnmowers in prairie restorations is the preferred method of control.
19. *Elaeagnus angustifolia*, Russian olive: This tree is not effectively controlled by the method of occasional prairie burns. Because Russian olive proliferates so efficiently and grows rapidly, it is best to cut the tree and paint the stump with undiluted glyphosate concentrate, 18–41% glyphosate.
20. *Lespedeza cuneata*, Asian lespedeza: Lespedeza is especially troublesome mainly because it is difficult to see in the usual mixed prairie situation. Overlooking a few plants can be disastrous, as it readily spreads by seed. When spotted, these plants should be pulled. Because of the plant often breaks away from the taproot, a dandelion spade should be utilized. Dense patches are easily removed by spraying with broadleaf herbicide, to the detriment of surrounding broadleaf natives.
21. *Miscanthus* species, silvergrass: There are three alien ornamental grasses that are beginning to spread from suburban lawns and biofuel projects, namely *Miscanthus sinensis*, *M. sacchariflorus*, and *M. floridulus*. I have experienced spread of *Miscanthus* from a suburban lawn (completely devoid of native species). In my hands glyphosate herbicide (1–2%) is effective in eliminating the problem.
22. Other alien species: Certain alien species can become potentially damaging as they move into new territories or adapt to new conditions. Managers of natural areas should be vigilant to the sudden appearance of unfamiliar plants. Some of these aliens are *Carduus nutans*, musk thistle (widespread); *Cirsium vulgare*, bull thistle (widespread); *Galega officinalis*, goatsrue (not the native goatsrue); *Heracleum mantegazzianum*, giant hogweed; *Microstegium vimineum*, Japanese stilt grass (mainly in shaded woodlands); *Polygonum cuspidatum*, Japanese knotweed (widespread); *Sorghum bicolor*, shattercane (widespread); *Sorghum halepense*, Johnson grass (widespread); *Acer platanoides*, Norway maple; *Ailanthus altissima*, tree of heaven (tree el diablo in my opinion); *Celastrus*

orbiculatus, oriental bittersweet; *Polygonum perfoliatum*, mile-a-minute vine (spreading rapidly from southeastern Pennsylvania); *Pueraria lobata*, kudzu (mainly in the South); *Aegopodium podagraria*, goutweed; *Bromus tectorum*, cheatgrass (widespread); *Perilla frutescens*, beefsteak plant; *Ranunculus ficaria*, fig buttercup/lesser celandine; *Arundo donax*, giant reed (mainly in the South); *Senecio jacobaea*, tansy ragwort (one plant produces up to 150,000 seeds, which can lie dormant for 20 years); *Akebia quinata*, fiveleaf akebia; *Ampelopsis brevipedunculata*, porcelain-berry; *Euphorbia esula*, leafy spurge (becoming widespread). Leafy spurge is particularly difficult to control, making early treatment of initial infestations important. Some managers consider the common *Trifolium pratense*, red clover, to be problematic, but I have observed its demise as the restored prairie matures.

Other Recommended Readings

Several other works regarding prairie restoration are available and afford additional excellent tips on successful reconstruction of native prairies. Kurtz [34] outlines useful methods for restoration of large acreages using farm equipment. Included are methods of using a modified combine to collect seed, spraying herbicide, using a harrow and soil compacter, broadcasting seeds with a dry fertilizer spreader, and postplanting mowing.

Fromm [35] explains how weed competition is kept to a minimum. He advocates soil preparation by multiple herbicide treatments followed by a no-till fall sowing of forbs seed. No-till prevents weed seed from being brought to the surface to promote its germination. Late in the fall of the year after forbs emerge, the prairie grasses are sown.

An excellent book edited by Packard and Mutel [10] gives a wealth of information on many topics. For example, additional data can be obtained regarding seed treatment, times for seed collection, weed control, seed mixes for various soil types, burn methods, herbicides, prairie wildlife, plant conservatism, organisms of the rhizosphere, and an extensive list of species. This work is highly recommended as a reference source written by leading experts in the field of prairie restoration. A very limited number of hand-drawn plants is the only drawback for the amateur, but undoubtedly not for prairie experts.

A prairie restoration book by Shirley [36] describes 104 species centered in Iowa. Only a few species are restricted to the eastern shortgrass Great Plains. This work is recommended for the amateur regarding its explicit detail on how to restore a prairie. Line-drawn illustrations are provided for all species in flower and with seed, but many drawings would be difficult as a guide to identify the plant for anyone unfamiliar with prairie plants. The methods of restoration are thoroughly described with a few deficiencies in emphasis. In my opinion, undue stress was placed on growing potted plants for placing outdoors, which is a labor-intensive procedure that should be restricted to only the most difficult conservative species. Also, the value of no-till seeding on glyphosate-treated soil was not sufficiently emphasized, as well as herbicide treatment in general. Only a few recommendations for seed germination were found to be unsuitable.

H.W. Gardner, *Tallgrass Prairie Restoration in the Midwestern and Eastern United States: A Hands-On Guide*, DOI 10.1007/978-1-4419-7427-3,

Appendix 1
Plants of the Tallgrass Prairie

Plant list is applicable to the tallgrass region centered in Illinois and can apply to comparable climate zones in the East (Plant Hardiness Zones 5 and 6); species composition varies somewhat outside of these hardiness zones and available rainfall. Compiled from various sources: Taft et al. [6], Packard and Mutel [10], Prairie Moon Nursery Catalogs [11], Deno [12], Flora of the Great Plains [15], Gray's Manual of Botany [14], Britton and Brown [13], Swink and Wilhelm [16], R.H. Mohlenbrock [17, 19, 22, 37–43], A.S. Hitchcock [44], J.R. Heckert [45], http://plants.usda.gov/java/nameSearch, and H.W. Gardner. The number of varieties (var., v.), subspecies (ssp.), and forms are a total for the species; all are not necessarily found in the tallgrass prairie area. A listing of one variety indicates that other varieties are not found or are alien varieties. Species found on other continents are considered to be alien species and are not included (whether justified or not). Many wetland species typical of wet prairies are also included; see OBL under the WET key. Listing is often subjective; that is, weedy species are omitted and certain rare or attractive full-sun-intolerant plants are included.

Key

Asterisk*

Indicates plants most commonly found on average dry-mesic/mesic/wet-mesic prairie or savanna.

H_2O

Moisture regime of growth preference. 1 = very wet; 5 = very dry.

Eco

Preferred ecosystem. P=prairie (full sun); S=savanna (sparse trees, partial shade); W=woodland (full canopy); N=well-drained soil, such as sand, gravelly soil, or friable black loam; for some species the preference is more stringent than for others.

Seed

Seed treatment needed for germination is largely based on Prairie Moon Nursery [11], except where superseded by detailed germination experiments by Deno [12]. (:=and treatment, /=or treatment). Reasonably successful fall planting/sowing in the author's experience is also listed. If seed treatment is not listed, try treatment listed for other members of the genus.

A=No pretreatment necessary.

C=Moist, cold stratification is needed. Mix seeds with equal or a larger amount of damp sand (not dripping wet), and put in a slightly open plastic bag for refrigerated storage (not freezer). Two months of storage are necessary, with exceptions in days listed with C; for example C90 means 90 days of stratification.

D=Seeds are very small or need light to germinate. Do not cover seeds after sowing. However, a single layer of burlap or cotton sheet can be used to prevent the soil from drying out (or shade with a window screen). Sowing in areas containing plant matter is also satisfactory.

E=Seeds need warm-moist followed by cold-moist treatment. Store in damp sand (in plastic bag slightly open) in warm place for 60–90 days then store in refrigerator for 60–90 days. Alternative is to sow outdoors and allow a full year for germination.

F=Soak seeds in water for 24 h; drain off excess water and store in sealed plastic bag at room temperature for 30–60 days. Then store in refrigerator for 45–60 days.

G=Sow seeds in late fall or early spring as these plants like coolness.

H=Seeds need scarification. Rub seeds between two sheets of medium grit sandpaper before moist-cold stratification (C). Deno [12] reports that a hole must be made in the seed coat.

I=Legume seeds need to be inoculated with nitrifying bacteria specific to each species. The author found that this treatment is generally unnecessary, except *Dalea* species seem to thrive better.

J=Seeds that have a hull or other covering should be removed as described in treatment H or by other methods.

K=Parasitic plants needing a host, such as hairy grama, blue grama, buffalo grass, penn sedge, June grass, or little bluestem. Cut a slight crevice at the base of the host. Sow seed in the cut. Alternative is to sow host and parasitic seeds together. *Aureolaria* and *Gerardia* are hemiparasitic on oak roots. *Orobanche* species are parasitic on composites.

L=Plant fresh seed.

M=Best fall planted.

N=Dry store seeds for 6 months at 40ºF.

O=Dry store seeds for 6 months at 70ºF.

P = Plant/sow outdoors soon after seed collection.
Q = Seeds may require 2 years or more to germinate or appear above ground.
R = Seeds washed three times daily for 1 week.
S = Temperatures above about 80 to 90°F are needed for germination.

Wet

Wetland code: National wetland categories for North Central Region according to US Fish and Wildlife Service [Reed 1988, see ref. Taft et al. [6]]. OBL: Obligate wetland (>99% probability of being found in a wetland). FACW: Facultative wetland (occasionally found in non-wetlands; 67–99% probability of being found in wetland). FAC: Equally likely to occur in wetlands or non-wetlands (34–66% probability of being found in a wetland). FACU: Facultative upland (occasionally found in wetlands, but usually occurs in non-wetlands with a 1–33% probability of being found in a wetland). UPL: Upland (almost always not found in wetland with <1% probability of being found in a wetland).

A positive sign (+) indicates a tendency toward a higher frequency of wetness; a negative sign (–) indicates less frequency of wetlands in the indicated category.

STAT/CC

Listing STATus; E = Endangered in Illinois. T = Threatened in Illinois. F = federal list. CC: Coefficient of conservatism [6] as indicator of degree of disturbance; 10 indicates those species least tolerant of disturbance (high-quality prairie or highest order of plant succession); 0 indicates species most tolerant of disturbance (weedy tendency); * indicates native species introduced into Illinois from neighboring areas. See Taft et al. [6].

Seed Time

Optimum time to collect seeds in Plant Hardiness Zones 5 and 6.

FLR Time

Early or peak time of flowering in Plant Hardiness Zones 5 and 6.

PREDTN

Susceptibility to predation. D, deer (or other herbivore); P, plant; S, seed or seedhead; F, flower; M, mice or other rodent; R, root; C, root corm; L, larva/maggot; I, insect including caterpillar; *good potential to recover.

	H_2O	Eco	Seed	Wet	STAT/CC	Seed-time	FLR-time	PREDTN
Agalinis aspera, rough false foxglove	4–5	P, S, N		UPL	10			
Agalinis auriculata, earleaf gerardia (annual)	3–4	P, S		UPL	T/8			
Agalinus fasciculata, fascicled agalinis	2–3	P, S		FACW	6			
Agalinis gattingeri, Gattinger's agalinis	4–5	P, S, W, N		UPL	10			
Agalinis paupercula, small-flowered agalinis (2 var.)	1–2	P, S		OBL	7		9/2	
Agalinis purpurea, large purple agalinis	2–3	P, S		FACW	6		9/12	
Agalinis skinneriana, Skinner's agalinis	4–5	P, S, N		UPL	9		9/1	
Agalinis tenufolia, slender false foxglove (5 var.)	2–4	P, S	C:K:D	FACW	5		9/1	
Agastache foeniculum, blue giant/anise hyssop	2–4	P, S	C:D/A:D	–	–	9/1–		
Agastache nepetoides, yellow giant hyssop	2–4	P, S	C90:D/A:D	FACU	4		8/1	
Agastache scrophulariifolia, purple giant hyssop	2–3	P, S	O:D/M:D	UPL	5		8/15	
Agrimonia parviflora, small/many-flowered agrimony	2	P, S		FAC+	5		8/15	
Agropyron see *Elymus trachycaulis*								
Agrostis elliottiana, awned bent grass	4–5	P, S		UPL	5			
Agrostis hyemalis, hair/tickle grass	2–4	P, S		FAC–	2		5/1	
Agrostis scabra, rough bent/tickle grass	2–4	P, S		FAC	5			
Aletris farinosa, colic root	2–3	P, N	A:D	FAC	9		7/5	
**Allium canadense*, wild garlic (6 var.)	2–4	P, S, W	A	FACU	2	6/19 bulbs	6/16	
**Allium cernuum*, nodding wild onion (3 var.)	2–4	P, S	A/M	UPL	7	9/1–9/25	7/19	DP*
Allium stellatum, prairie onion	3–5	P	C/M	UPL	10	9/18–10/3	8/8	DP*
Ammannia coccinea, long-leaved ammannia	1–2	P, S		OBL	5		8/20	
**Amorpha canescens*, lead plant	3–5	P, S, N	C10:J	UPL	8	9/5–10/17	7/5	
Amorpha fruticosa, false indigo bush (many var.)	2–3	P, S	H:J	FACW+	6		6/5	
Amphicarpaea bracteata, hog peanut	3	P, S		FAC	4			
Amsonia tabernaemontana, blue dogbane (3 var.)	2–3	P, S	C90/M	FACW	6	9/4–10/3	5/16	
Anagalis minima, chaffweed	1–2	P, N		–	–			
Andropogon also see *Schizachyrium*								
**Andropogon gerardii*, big bluestem	2–5	P	A/M	FAC–	5	9/26–10/30	7/19	
Andropogon gyrans, Elliott's broom sedge	4–5	P		UPL	3			

Andropogon ternarius, splitbeard bluestem	2–4	P, S		FACU	8			
Andropogon virginicus, broom sedge (2 var.)	2–5	P, S		FAC–	1	9/1–10/12		
Androsace occidentalis, androsace	4–5	P, S, N	FACU–	4			4/25	
**Anemone canadensis*, Canada anemone	2–3	P, S	F	FACW	4	7/13–8/20	5/16	
Anemone caroliniana, Carolina anemone	4–5	P, N	C	UPL	9		4/30	
**Anemone cylindrica*, thimbleweed	3–5	P, S	N/O/C90/M	UPL	8		6/27	DP*
Anemone patens see *Pulsatilla patens*								
**Anemone virginiana*, tall anemone (3 var.)	3–4	P, S, W	C, G/M	UPL	4	9/4–10/22	6/25	DP*
Angelica atropurpurea, angelica	2	S	F/M	OBL	6	6/26–7/18	6/9	
**Antennaria neglecta*, pussytoes	4–5	P, S	C:D	UPL		5/30	4/25	
Antennaria howellii ssp. *neodioica*, pussytoes	3–5	P, S		UPL	*	5/30	4/28	
Antennaria plantaginifolia, everlasting	4–5	P, S	C:D	UPL	4	6/1–	4/25	
Apocynum androsaemifolium, spreading dogbane	3–5	P, S		UPL	6		6/15	
**Apocynum cannabinum*, dogbane	2–4	P, S		FAC	2	9/16–11/21	6/30	
Apocynum × *floribundum*, intermediate dogbane	2–4	P, S		UPL	6			
Arabis drummondii, Drummond's rock cress	2–4	P, S		FACU	10			
Arabis glabra, tower mustard	3–4	P		UPL	6		6/10	
Arabis lyrata, lyre-leaved rock cress	4–5	P, S, N		FACU–	7		4/25	
Aristida basiramea, fork-tipped three-awn grass	4–5	P, S		UPL	6			
Aristida dichotoma, poverty grass	3–5	P, N		FACU	2			
Aristida longespica, 3-awn (2 var.)	3–5	P, N		FACU–	2			
Aristida purpurascens, arrowfeather (3 var.)	4–5	P, N		UPL	5			
Aristida ramosissima, slender three awn	3–5	P		UPL	3			
Aristida tuberculosa, needle grass	3–5	P, S, N		UPL	9			
**Arnoglossum atriplicifolium*, pale Indian plantain	3–4	P, S	C/M	UPL	5	9/15–10/3	7/4	
**Arnoglossum plantagineum*, prairie Indian plantain	2–3	P, N	C/M	FAC	10	7/26–9/10	7/7	
Artemisia campestris, beach wormwood (2 ssp.)	3–5	P, S, N		UPL	4		9/1	
Artemisia dracunculus, silky wormwood	4–5	P, N		UPL	T/9			

(continued)

(continued)

	H_2O	Eco	Seed	Wet	STAT/CC	Seed-time	FLR-time	PREDTN
Artemisia ludoviciana, Louisiana sage (8 ssp.)	4–5	P, S	A:D/C:D	UPL	2	10/28–11/7	7/4	
Artemisia serrata, saw-toothed sage	4–5	P, S	A:D	UPL	10			
Asclepias amplexicaulis, sand milkweed	4–5	P, N	C	UPL	7		6/20	
Asclepias hirtella, hairy green milkweed	4–5	P, N	C/A	UPL	6		7/26	
**Asclepias incarnata*, swamp milkweed (2 ssp.)	1–3	P	C90:D	OBL	4	9/16–10/1	7/20	
Asclepias lanuginosa, wooly milkweed	4–5	P, N		UPL	E/10		6/14	
Asclepias meadii, Mead's milkweed	4–5	P, N		UPL	E, FT/10			
Asclepias ovalifolia, oval milkweed	3–4	P, S		UPL	E/10			
**Asclepias purpurascens*, purple milkweed	2–4	P, S, N	C/A/M	FACU	7	9/9–9/18	6/19	DP*
Asclepias stenophylla, narrow-leaved green milkweed	4–5	P, N		UPL	T/10			
**Asclepias sullivantii*, prairie milkweed	2–3	P, N	C/A	UPL	7	8/26–10/17	7/9	I*
**Asclepias tuberosa*, butterfly weed (2 ssp.)	3–5	P, S, N	C90/M	UPL	5	8/18–9/15	7/9	I*
**Asclepias verticillata*, horsetail milkweed	3–5	P, S	C/A/M	UPL	1		8/1	
**Asclepias viridiflora*, green milkweed	4–5	P, N	C90/M	UPL	9	8/25–10/15	7/5	I*
Asclepias viridis, green-flowered milkweed	4–5	P, S, N	C/A	UPL	6			
Aster see *Symphyotrichum*, *Doellingeria*, and *Oligoneuron album*								
**Astragalus canadensis*, Canadian milk vetch	2–4	P, S	C10:H/M	FAC+	7	7/20–9/5	7/19	
Astragalus crassicarpus, large ground plum (5 var.)	4–5	P, S, N	C10:H	UPL	E/8			
Astragalus distortus, bent milk vetch (2 var.)	4–5	P, S, N		UPL	8			
Astragalus tennesseensis, Tennessee milk vetch	4,5	P, S		UPL	E/10			
Aureolaria grandiflora, false foxglove (4 var.)	3–5	S	C:K:D/M:K	UPL	6	10/18–11/1	8/12	
**Baptisia alba* v. *macrophylla*, white wild indigo	2–5	P, S	C10:H/M:Q	FACU	6	8/27–9/25	6/25	DS/IP*
Baptisia australis, blue wild indigo (3 var.)	2–3	P, S	L:A/O:H	UPL	*	7/17–8/1	5/25	
**Baptisia bracteata* v. *leucophaea*, cream wild indigo	3–5	P, S	C10:H/M:Q	UPL	9	8/8–9/1	5/16	DS*
Baptisia tinctoria, yellow wild indigo	3–5	P, S, N	C10:H/M	UPL	10	8/19–9/10	7/2	
Beckmannia syzigachne, American slough grass	1–2	P	A	OBL	E/10		7/21	
Berula erecta, water parsnip	1–2	P		OBL	10		8/12	
Besseya bullii, kittentail	4–5	P, S, N	M/C120:D	UPL	T/8			

Bidens aristosa, swamp marigold	1–2	P, S	C/A/M	FACW	1	9/18–9/30	8/9
Bidens cernua, nodding bur marigold	1–2	P	C	OBL	2		8/10
Bidens connata, purple-stemmed tickseed	1–2	P	C	OBL	2		
Bidens coronata, tall swamp marigold	1–2	P		OBL	7		8/10
Bidens tripartita, leafy-bracted tickseed	2–3	P		–	–		9/22
Blephilia ciliata, downy wood mint	3–5	P, S, W	C:D	UPL	6	7/16–	6/5
Boltonia asteroides, false aster (3 var.)	1–2	P, S, N	C:D	FACW	5		9/15
Boltonia decurrens, decurrent false aster	1–2	P	C:D	OBL	T, FT/4		
Boltonia diffusa, doll's daisy	1–2	P, S		FACW	4		
**Bouteloua curtipendula*, sideoats grama (2 var.)	4–5	P, S	A/M	UPL	7	8/14–10/3	8/8
Bouteloua gracilis, blue grama	4–5	P, S, N	A/M	UPL	5		
Bouteloua hirsuta, hairy grama (2 var.)	5	P, S, N	A	UPL	9		
**Brickellia eupatorioides*, false boneset (6 var.)	4–5	P, S	A/M	UPL	6	8/30–10/15	8/27
Bromus ciliatus, fringed brome (2 var.)	1–2	P	A	OBL	10		7/29
Bromus kalmii, prairie brome	2–4	P, S, N	A	FAC	10		6/20
Bromus latiglumis, ear-leaved brome	2–3	P, S		FACW–	7		
Buchnera americana, blue hearts	2–3	P, S, N	K	FAC–	10		
Bulbostylis capillaris, hair sedge (2 ssp.)	1–2	P, S, N		FACU+	4		
Cacalia see *Arnoglossum*							
**Calamagrostis canadensis*, bluejoint grass (3 var.)	1–3	P	A:D	OBL	3		7/2
Calamagrostis stricta, bog reed grass (2 ssp.; 2 var.)	2	P		FACW+	5		6/30
Calamovilfa longifolia, sand reed (2 var.)	4–5	P	A	UPL	8		
Callirhoe alcaeoides, pale poppy mallow	4–5	P, S, N		UPL	5		
Callirhoe triangulata, poppy mallow	4–5	P, S, N	C:H/M	UPL	9		7/18
Calopogon tuberosus, grass pink orchid (2 var.)	1–2	P		OBL	T/10		7/7
Caltha palustris, marsh marigold	1–2	P, S		OBL	7		
Calystegia spithamaea, dwarf bindweed (3 ssp.)	3–5	P, S, N		UPL	10		6/15
Camassia angusta, wild hyacinth	3–4	P, S	C	UPL	E/7		

(continued)

(continued)

	H_2O	Eco	Seed	Wet	STAT/CC	Seed-time	FLR-time	PREDTN
**Camassia scilloides*, wild hyacinth	2–3	P, S	C30/M	FAC+	7	6/11–6/24	5/15	DPFS*
Campanula aparinoides, marsh bellflower	1–2	P, S		OBL	8		8/10	
Campanula rotundifolia, harebell	2–4	P, S	N/O/L:E	FAC–	8		6/23	
Cardamine bulbosa, bulbous cress	1–2	P, S		OBL	5			
Carex abdita, sedge	4–5	P, S, N		UPL	6			
Carex aggregata, glomerate sedge	3–5	P, S		UPL	4			
Carex alata, winged oval sedge	1–2	P, S		OBL	10			
Carex albolutescens, long-fruited oval sedge	2	P, N		FACW	9		5/5	
Carex alopecoidea, foxtail sedge	2–3	P	C:D	FACW+	4		6/5	
**Carex annectens*, yellow fox sedge	2–4	P		FACW	3		6/5	
Carex aquatilis, water sedge (3 var.)	1–2			OBL	6		5/14	
Carex arkansana, Arkansas sedge	3–5	P, S		FACU	8			
Carex atherodes, awned sedge	1–2	P		OBL	T/6		6/7	
Carex aurea, golden sedge	1–2	P, N		FACW+	9		6/2	
Carex bebbii, beautiful/Bebb's sedge	1–2	P	C:D	OBL	8		5/24	
**Carex bicknellii*, Bicknell's sedge	3–5	P, S	C:D	FAC–	8	9/16–	5/25	
**Carex brevior*, fescue sedge or hammer's sedge	3–4	P, S, N	C:D	FAC	4	6/17–6/29	5/15	
Carex bromoides, brome hummock sedge (2 ssp.)	1–2	P, S, W		OBL	10		4/25	
Carex bushii, Bush's sedge	3–4	P, S, W		FACW	4		5/29	
Carex buxbaumii, Buxbaum/brown sedge	1–2	P		OBL	9		5/14	
Carex caroliniana, Carolina sedge	3	P, S		FAC	7			
**Carex cephalophora*, capitate sedge	3–4	P, S		FACU	3			
Carex comosa, porcupine sedge	1–2	P	C:D	OBL	6		6/5	
**Carex conjuncta*, soft fox sedge	2	P, S, W		FACW	5		5/30	
Carex conoidea, sedge	1–2	P		OBL	10		5/22	
Carex crawei, Crawe's sedge	2–3	P, N		OBL	T/7		5/12	
**Carex cristatella*, crested sedge	1–2	P, S, W	C:D	FACW+	3		6/10	
Carex crus-corvi, raven's foot sedge	1–2	P, S		OBL	6			

Carex cryptolepis, small yellow sedge	1–2	P, N		OBL	9		5/14
**Carex davisii*, awned graceful sedge/Davis's sedge	2–3	P, S, W		FAC+	3	6/18–6/30	5/29
Carex diandra, bog panicled sedge	1–2	P, S		OBL	10		6/10
Carex emoryi, Emory's sedge	1–2	P		OBL	6		5/10
**Carex festucacea*, fescue sedge	2–3	P, S		FAC	6		
Carex frankii, Frank's sedge	1–2	P, S		OBL	4		
Carex garberi, false golden sedge	2–3	P, N		FACW	10		6/14
**Carex granularis*, pale/meadow sedge (2 var.)	2–3	P, S		FACW+	2		5/13
**Carex gravida*, heavy sedge (2 var.)	3–4	P	M:D	UPL	4	6/5–6/25	5/2
**Carex grayi*, Gray's sedge	1–2	P, S, W	O:D/A:D	FACW+	6		7/2
**Carex grisea*, wood gray sedge	3–4	P, S, W	C	UPL	3		5/28
Carex haydenii, Hayden's sedge	1–2	P, N		OBL	7		6/2
Carex heliophila, yellow sedge	3–4	P, S					
**Carex hirsutella*, hairy-leaved sedge	3–5	P, S, W	C:D	FACU–	5		5/30
Carex hyalinolepis, sedge	1–2	P		OBL	4		6/2
Carex hystericina, porcupine/bottlebrush sedge	1–2	P	C:D	OBL	6		6/20
Carex interior, inland sedge	1–2	P, S, W	C:D	OBL	8		5/20
Carex intumescens, shining bur sedge	1–2	P, S, W		FACW+	9		7/8
Carex lacustris, river/lake-bank sedge	1–2	P, S, W	C:D	OBL	6		6/21
Carex laeviconica, sedge	1–2	P, S	C:D	OBL	10		6/25
Carex lasiocarpa v. *americana*, hairy-fruited sedge	1–2	P		OBL	10	6/18–6/30	5/30
Carex leptalea, slender sedge (3 ssp.)	1–2	P		OBL	10		5/13
Carex longii, Long's sedge	2–3	P, S, N		FAC	8		5/15
**Carex lurida*, sallow sedge	1–2	P, S, W	J:D	OBL	7		6/9
Carex meadii, Mead's sedge	2–3	P		FACU–	6		5/11
**Carex mesochorea*, oval-headed/midland sedge	4	P, S					
Carex microrhyncha, sedge	4–5	P, S, N		–	–		4/20
**Carex molesta*, sedge	1–2	P, S	C:D	FAC	2		5/14

(continued)

(continued)

	H_2O	Eco	Seed	Wet	STAT/CC	Seed-time	FLR-time	PREDTN
**Carex muhlenbergii*, Muhlenburg's/sand sedge (2 var.)	5	P, S, N	C:D	UPL	5		6/10	
**Carex pellita*, woolly sedge	1–2	P		OBL	4		5/11	
Carex praegracilis, silver sedge	2–3	P		FACW	*		5/30	
Carex prairea, prairie sedge	1–2	P	C:D	OBL	10		6/4	
Carex projecta, loose-headed oval sedge	1–2	P, N		OBL	4		6/25	
Carex retrorsa, deflexed bottlebrush sedge	1–2	P	A:D/C:D	OBL	7		6/19	
Carex richardsonii, hummock sedge	4–5	P, N		UPL	10		5/10	
Carex sartwellii, Sartwell sedge (2 var.)	1–2	P		OBL	5		5/20	
**Carex scoparia*, pointed broom sedge (2 var.)	2–3	P, S	C:D	FACW	5		6/1	
**Carex shortiana*, Short's sedge	2–3	P, S, W	C:D	FACW+	4	6/24–7/12	6/1	
Carex siccata, silvertop/hay sedge (2 var.)	3–5	P, N		UPL	7		5/10	
Carex squarrosa, squarrose sedge	1–2	P, S, W	C:D	OBL	5		7/26	
Carex sterilis, sedge	1–2	P	C:D	OBL	10		5/10	
**Carex stipata*, prickly/awl-fruited sedge (2 var.)	1–2	P, S, W	C:D	OBL	2		6/15	
Carex stricta, tussock sedge	1–2	P, S	O:D	OBL	5		5/11	
Carex suberecta, prairie straw sedge	1–2	P		OBL	7		6/1	
Carex tetanica, common stiff sedge	1–2	P		FACW	5		5/16	
Carex tonsa, smooth-fruited oak sedge (2 var.)	4–5	P, S		UPL	8		4/20	
**Carex tribuloides*, blunt broom sedge (2 var.)	1–2	P, S, W	C:D	FACW+	3		6/10	
Carex trichocarpa, hairy-fruited sedge	1–2	P		OBL	6		7/14	
Carex utriculata, beaked sedge	1–2	P, S		OBL	9			
Carex vesicaria v. *monile*, blister/inflated sedge	2–3	P	C:D	OBL	9	6/10–6/30	5/13	
**Carex vulpinoidea*, fox sedge (2 var.)	1–3	P, S	C:D/M:D	OBL	3	6/25–7/15	5/21	
Cassia see *Chamaecrista* and *Senna*								
Castilleja coccinea, Indian paintbrush	2–3	P, N	C:K:D	FAC	8	6/16–	5/19	
Castilleja sessiliflora, downy painted cup	4–5	P, N	C:K:D	UPL	E/10		6/5	
**Ceanothus americanus*, New Jersey tea	3–5	P, S	J:M/J:C90	UPL	8	8/12–9/8	6/21	DPFS*
Ceanothus herbaceus, red root	4–5	P, N	H:C	UPL	E/9		6/7	

Cerastium arvense, prairie/field chickweed	3–4	P, S		FACU–	4			
**Chamaecrista fasciculata*, partridge pea (annual, 2 var.)	3–5	P, S	C10:H/M	FACU–	1	9/11–10/2	8/8	
Chamaecrista nictitans, wild sensitive plant (2 var.)	4–5	P, N		FACU–	2	9/18–9/30	8/30	
Chamaesyce geyeri, Geyer's spurge (2 var.)	4–5	P, N		UPL	10			
Chasmanthium latifolium, wild oats	2–3	P		FACW	4			
Chelone glabra, turtlehead	1–2	P, S	O:D	OBL	7		9/7	
Chelone obliqua, pink turtlehead (3 var.)	1–2	P, S		OBL	8			
Chloris verticillata, windmill grass	3–4	P		UPL	*			
Cicuta bulbifera, bulblet-bearing water hemlock	1–2	P		OBL	9		8/10	
Cicuta maculata, water hemlock (4 var.)	1–2	P, S	M/C120	OBL	4	8/1–	7/4	
Cirsium carolinianum, Carolina thistle	4–5	S		UPL	8			
**Cirsium discolor*, field thistle	2–3	P, S		UPL	3		8/29	
Cirsium hillii, Hill's thistle	4–5	P, S, N	C	UPL	7		7/5	
Cirsium muticum, swamp thistle	1–2	P		OBL	9		9/9	
Cladium mariscoides, twig rush	1–2	P, S		OBL	10			
Clinopodium arkansanum, low calamint	1–2	P, S		FACW	8			
**Comandra umbellata/richardsiana*, false toadflax	2–4	P, S, N	K	FACU	6	7/1–	5/5	
Commelina erecta, narrow-leaved day flower (3 var.)	4–5	P, N		UPL	5		8/1	
Cordalis aurea, golden cordalis	2–4	P, S, N						
Cordalis micrantha, small-flowered cordalis	2–4	P, S, N						
**Coreopsis lanceolata*, lance-leaved coreopsis	4–5	P, N	O/M	FACU	5	6/24–8/10	6/16	
**Coreopsis palmata*, prairie coreopsis	3–5	P, S	C/M	UPL	6	9/18–10/3	6/25	
**Coreopsis tripteris*, tall coreopsis	2–4	P, S	C/M	FAC	4	10/8–10/23	9/6	DP*
Crotalaria sagittalis, rattlebox (annual, 2 var.)	4–5	P, S, N	A	UPL	3		8/15	
Croton glandulosus v. *septentrionalis*, sand croton	4–5	P, N		UPL	1		7/20	
Croton monanthogynus, croton	4–5	P, N		UPL	2		9/15	
Cunila origanoides, dittany	3–5	P, S		UPL	5			
Cuphea viscosissima, blue waxweed	3–5	P	A	FACU	4		8/26	

(continued)

(continued)

	H_2O	Eco	Seed	Wet	STAT/CC	Seed-time	FLR-time	PREDTN
Cuscuta cuspidata, cuspidate dodder	2–3	P	FACW+	5				
Cuscuta glomerata, glomerate dodder	2–4	P		FAC	6		8/25	
Cuscuta gronovii, Gronovius' dodder (2 var.)	2–3	P		FACW	2		8/18	
Cuscuta indecora, false field dodder (4 var.)	2–3	P, S, N		FAC	5			
Cuscuta pentagona, prairie dodder (3 var.)	4–5	P	C	UPL	5			
Cuscuta polygonorum, smartweed dodder	3–5	P		UPL	5			
Cyperus acuminatus, short-pointed cyperus	1–2	P		OBL	2			
Cyperus echinatus, hedgehog club rush	3–4	P, S, N		FAC	2			
Cyperus erythrorhizos, red-rooted cyperus	1–2	P, N		OBL	1			
Cyperus grayioides, umbrella/Gray's sedge	4–5	P, N		UPL	T/8			
Cyperus houghtonii, Houghton's cyperus	4–5	P, N		UPL	7			
Cyperus lancastriensis, Lancaster umbrella sedge	2–3	P, S, N		FAC–	7			
Cyperus lupulinus, great plains flatsedge (2 ssp.)	4–5	P, N	C:D	FACU–	5			
Cyperus odoratus, Engelman's cyperus	1–2	P		OBL	7			
Cyperus pseudovegetus, marsh cyperus	1–2	P, S		FACW	5			
Cyperus schweinitzii, Schweinitz flatsedge	4–5	P, S, N	C:D	FACU+	5			
Cyperus squarrosus, bearded flat sedge	1–2	P		OBL	2			
Cypripedium candidum, white lady's-slipper orchid	1–2	P		OBL	E/10		5/13	
Cypripedium parviflorum, yellow lady's slipper (3 var.)	2–4	S		FAC+	8		5/19	
Cypripedium reginae, showy lady's slipper	1–2	P, S		FACW+	E/10		6/15	
**Dalea candida*, white prairie clover (2 var.)	3–4	P	A:J:I/M	UPL	9	9/4–10/15	7/9	MorDP*
Dalea foliosa, leafy prairie clover	3–4	P	A:J:H:I/M	UPL	E, FE/10	8/21–10/17	7/14	MorDP/ MR
**Dalea purpurea*, purple prairie clover (2 var.)	3–5	P	A:J:I/M	UPL	8	8/21–10/8	7/14	MorDP/ MR
Danthonia spicata, poverty oat grass	3–5	P, S		UPL	3			
Dasiphora fruticosa, shrubby cinquefoil	2–4	P		FACW	10			
Daucus pusillus, small wild carrot or rattlesnake weed	3–5	P, S, N		FACU–	E/*			

Decodon verticillatus, swamp loosestrife	1–2	P, S		OBL	8		8/14	
Delphinium carolinianum, wild blue larkspur (3 ssp..)	4–5	P, N	C:G	UPL	10	6/24–7/20	6/24	
Deschampsia cespitosa, tufted hair grass	2	P		FACW+	8		7/4	
**Desmanthus illinoensis*, Illinois bundleflower	2–4	P	A:H/M	FAC–	4	8/29–10/17	7/19	DF*
**Desmodium canadense*, showy tick trefoil	2–4	P	A:J:H/M	FAC–	5	9/5–9/18	8/8	
Desmodium canescens, hoary tick trefoil	3–5	P, S, N	A:J	UPL	4			
Desmodium ciliare, hairy tick trefoil (2 var.)	4–5	P, S, N		UPL	7			
**Desmodium illinoense*, Illinois tick trefoil	4–5	P, S	A:J	UPL	5	8/1–	7/15	
Desmodium laevigatum, glaucous tick trefoil	3–4	P, S, N		FACU				
Desmodium marilandicum, smooth small-leaved tick trefoil	4–5	P, S, N		UPL	6			
Desmodium obtusum, stiff tick trefoil	3–4	P, S, N		FACU	5			
Desmodium paniculatum, panicled tick trefoil (2 var.)	3–4	P, S	A:J	FACU	2			
Desmodium sessilifolium, sessile-leaved ticktref	4–5	P, S, N	A:J	UPL	6		8/15	
Dichanthelium acuminatum, panic grass (6 var.)	2–4	P, S	A	FAC	2			
Dichanthelium boreale, northern panic grass	2–3	P		FACU+	10			
Dichanthelium depauperatum, panic grass	4–5	P, S		UPL	7			
Dichanthelium leibergii, Leiberg's panic grass	3–5	P, S, N		FACU+	7			
Dichanthelium linearfolium, slimleaf panic grass	4–5	P, S, N		UPL	7			
**Dichanthelium oligosanthes* (2 var.)	3–5	P, S, N	A	FACU	3	6/18–6/28	5/25	
Dichanthelium scoparium, velvety panic grass	2–3	P, S		FACW	6			
Dichanthelium villosissimum, panic grass (2 var.)	4–5	P, S, N		UPL	5			
Dichanthelium wilcoxianum, panic grass	3–5	P, S, N		UPL	10			
Digitaria cognata, fall witch grass	4–5	P, N		UPL	4		8/13	
Digitaria filiformis, slender crab grass	4–5	P, N		UPL	4		9/1	
Diodia teres, poorjoe or rough buttonweed (5 var.)	4–5	P, N		FACU	2		8/18	
Dodecatheon amethystinum, amethyst shooting star	2–4	P, S, W	D:G:C/M	UPL	E/9	6/29–7/13	5/1	
**Dodecatheon meadia*, shooting star (2 ssp.)	2–4	P, S	M/C21:D:G	FACU	6	7/3–7/17	5/9	DF*
Doellingeria umbellata, flat-top aster (2 var.)	2–3	P, S	C	FACW	8		9/1	

(continued)

(continued)

	H_2O	Eco	Seed	Wet	STAT/CC	Seed-time	FLR-time	PREDTN
Draba brachycarpa, short-fruited Whitlow grass	3–4	P, S		UPL	2			
Draba reptans, Whitlow grass (2 var.)	4–5	P, S, N		UPL	3		5/1	
Drosera intermedia, narrow-leaved sundew	1–2	P, N		OBL	T/10	7/30		
**Echinacea pallida*, pale purple coneflower	3–5	P, N	D:C90/M	UPL	7	8/27–10/30	6/25	MR (*?)
**Echinacea purpurea*, purple coneflower	2–3	S	O/M	UPL	6	8/27–10/8	7/4	
Echinochloa walteri, salt-marsh cockspur grass	1–2	P		OBL	5			
Eleocharis acicularis, needle spike rush (2 var.)	1–2	P	C:D	OBL	3			
Eleocharis compressa, flat-stemmed spike rush	1–2	P		OBL	7			
Eleocharis elliptica, flat-stemmed spike rush (2 var.)	1–2	P		OBL	8		5/13	
Eleocharis erythropoda, red-rooted spike rush	1–2	P		OBL	3			
Eleocharis intermedia, matted spike rush	1–2	P		FACW	7			
**Eleocharis obtusa*, blunt spike rush	1–2	P		OBL	2		6/23	
Eleocharis olivacea, bright green spike rush (2 var.)	1–2	P		OBL	10			
Eleocharis palustris, great spike rush (2 var.)	1–2	P	C:D	OBL	8			
Eleocharis quinqueflora, matted spike rush	1–2	P		OBL	10			
Eleocharis rostellata, wicket spike rush	1–2	P		OBL	10			
Eleocharis tenuis, slender spike rush (3 var.)	1–2	P		OBL	7			
Eleocharis wolfii, Wolf's spikesedge	1–2	P		OBL	9			
**Elymus canadensis*, Canada wild rye	3–5	P, S	A/M	FAC–	4	8/25–10/17	7/3	
Elymus trachycaulus, bearded wheat grass (3 ssp.)	2–4	P, S	A	FAC	8			
**Elymus villosus*, hairy/slender wild rye	2–4	P, S, W	A/M	FACU	4	7/25–8/15	7/12	
**Elymus virginicus*, Virginia wild rye (2 var.)	2–3	P, S, W	A/M	FACW–	4	8/18–9/10	7/15	
Epilobium coloratum, cinnamon willow herb	1–2	P		OBL	3		8/16	
Epilobium leptophyllum, fen willow herb (2 form)	1–2	P		OBL	9		8/15	
Eragrostis frankii, sandbar love grass	4–5	P, N		FACW	2			
Eragrostis hypnoides, pony grass	2–3	P, N		OBL	5			
Eragrostis spectabilis, purple love/tumble grass	3–5	P, N	A:D	UPL	3			
Eragrostis trichodes, sand lovegrass (2 var.)	4–5	P, S, N	A:D	UPL	5			

**Erigeron philadelphicus*, marsh fleabane (2 var.)	2–3	P, S		FACW	3		6/5	
Erigeron pulchellus, poor robin's plantain (3 var.)	3–4	P, S	C:D	FACU	5		5/20	
**Erigeron strigosus*, daisy fleabane (5 var.)	2–4	P		FAC–	2		6/5	
Eriophorum angustifolium, narrow-leaved cotton grass (2 ssp.)	1–2	P	A:D/M:D	OBL	10			
Eriophorum gracile, slender cotton grass (2 var.)	1–2	P		OBL	10			
Eriophorum virginicum, rusty cotton grass	1–2	P		OBL	10			
Eriophorum viridi-carinatum, tall cotton grass	1–2	P		OBL	10			
Eryngium prostratum, eryngo	1–2	P, N		OBL	5			
**Eryngium yuccifolium*, rattlesnake master (2 var.)	2–4	P, S	M/C90:G	FAC+	7	9/25–10/29	7/4	MR*
Erysimum capitatum, western wallflower (6 var.)	4–5	P, N		UPL	7			
Erythronium mesochoreum, prairie dogtooth violet	3–5	P, S, W		UPL	9			
Eupatoriadelphus fistulosus, Joe Pye weed	1–2	P, S		OBL	7			
Eupatoriadelphus maculatus, Joe Pye (3 var.)	1–2	P, S	A:D/C:D	OBL	5	9/5–10/2	7/20	
**Eupatorium altissimum*, tall boneset	4	P, S	C/M	FACU	2	9/1–10/25	8/13	
**Eupatorium perfoliatum*, common boneset (2 var.)	1–2	P, S	O:D	FACW+	4	8/28–11/5	8/20	
Eupatorium purpureum, sweet Joe Pye weed (2 var.)	2–4	P, S	C90:D/M:D	FAC	5	8/6–9/15	7/30	
**Euphorbia corollata*, flowering spurge	3–5	P, S	C30/M	UPL	3	8/1–9/26	8/1	DP*
**Euthamia graminifolia*, grassleaf goldenrod (2 var.)	1–4	P, S	C:D	FACW–	3	11/9	9/1	
Euthamia gymnospermoides, *viscid grass-leaved goldenrod*	2–3	P, S		FAC+	5			
Festuca paradoxa, greater nodding fescue	2–4	P, S		FAC	6			
Filipendula rubra, queen-of-the-prairie	1–2	P, N	C90	FACW+	T/10	9/5–9/26	6/25	
Fimbristylis autumnalis, autumn sedge	2–4	P, S, N		FACW+	6			
Fimbristylis puberula, chestnut sedge (2 var.)	4–5	P, N		UPL	9			
**Fragaria virginiana*, wild strawberry (4 ssp.)	2–5	P, S	A	FAC–	2		5/7	
Gaillardia aestivalis, perrenial gaillardia	4–5	P		UPL	10			
Galium asperellum, rough bedstraw	1–2	P, S		OBL	7		8/16	
Galium boreale, northern bedstraw	2–4	P, S, N	A	FAC	7	8/1–	6/17	

(continued)

(continued)

	H_2O	Eco	Seed	Wet	STAT/CC	Seed-time	FLR-time	PREDTN
Galium obtusum, wild madder (3 ssp.)	1–2	P, S		FACW+	5		6/20	
**Galium tinctorium*, stiff marsh bedstraw	1–2	P, S		OBL	6		6/22	
Galium trifidum, small bedstraw (4 ssp.)	1–2	P		FACW+	10		8/1	
Galium virgatum, dwarf bedstraw	4–5	P, S, W		UPL	10			
Gamochaeta purpurea, early cudweed	3–5	P, S, N		FACU	2			
**Gaura biennis*, gaura/butterfly weed	2–5	P, S	M	FACU–	2	9/13–10/2	8/21	
Gaura longiflora, large-flowered gaura	3–5	P, S, N	C	UPL	5			
Gaura mollis, small-flowered gaura	4–5	P		UPL	*		7/22	
**Gentiana alba*, cream gentian	2–3	P, S	M:D/A:D	FACU	9	10/8–11/13	8/20	LS
**Gentiana andrewsii*, bottle gentian (2 var.)	2–3	P, S	C:D	FACW	7	10/8–10/30	8/19	
Gentiana clausa, closed gentian	2	P		FACW+	10			
**Gentiana puberulenta/puberula*, prairie/downy gentian	3–4	P	M:D/C90:D	FACU	9	10/15–11/6	9/6	LS
Gentiana saponaria, soapwort gentian (2 var.)	2–3	P, S, N	E:D	FACW–	9		9/25	
Gentianella quinquefolia, stiff gentian (2 ssp.)	2–4	P, S	C:D	FAC	7		9/20	
Gentianopsis crinita, fringed gentian	2	P, S, W	M:D	FACW+	10		9/15	
Gentianopsis virgata, small fringed gentian (2 ssp.)	2	P, N		OBL	10			
Geranium carolinianum, Carolina cranesbill (2 var.)	4–5	P, S, N		UPL	2		6/5	
**Geranium maculatum*, wild geranium	3–4	S, W	L	FACU	4	6/16–	5/4	
Geum aleppicum, yellow avens	2–3	P, S	C	FAC+	6		7/10	
Geum triflorum, prairie smoke/prairie avens (4 var.)	2–5	P, S	C	FACU–	9	5/15–	4/10	
Glandularia canadensis, rose vervain	4–5	P, N		UPL	7			
Glyceria septentrionalis, floating manna grass	1	P		OBL	6		6/14	
Glyceria striata, fowl manna grass	1–2	P, S	A:D	OBL	4		6/16	
Gnaphalium see *Pseudognaphalium*								
Hasteola suaveolens, sweet Indian plantain	1–2	P, S, W	C	OBL	10		9/12	
**Helenium autumnale*, autumn sneezeweed (3 var.)	1–2	P, S	A:D	FACW+	3		8/28	
Helenium flexuosum, purple-headed sneezeweed	1–3	P, S	A	FAC+	4		7/20	
Helianthemum bicknellii, frostweed/rockrose	4–5	P, S, N		UPL	7	7/1–	6/20	

Helianthemum canadense, common rockrose	4–5	P, S, N		UPL	7	7/16–	6/19	
Helianthus angustifolius, narrow-leaved sunflower	2–3	P, S		FACW–	10			
Helianthus decapetalus, pale sunflower	3–5	S		UPL	5			
Helianthus divaricatus, woodland sunflower	3–5	P, S	C/A	UPL	5		8/20	
Helianthus × doronicoides, oblong-leaved sunflower	3–5	P, S		–	–			
Helianthus giganteus, tall/giant wild sunflower	2	P, S	C/A	FACW	E/9		8/27	
Helianthus grosseserratus, sawtooth sunflower	2–4	P, S	C/A	FACW–	2	9/1–9/30	9/7	DP*
Helianthus hirsutus, oblong sunflower (3 var.)	3–5	S		UPL	5			
**Helianthus × laetiflorus*, showy sunflower (*H. rigidus × H. tuberosus*)	4–5	P, S	C/A	UPL	*	9/7–10/1	9/3	
Helianthus maximiliani, Maximillian's sunflower	2–4	P	C/A/M	UPL	*	10/21–10/30	9/21	
**Helianthus mollis*, hairy sunflower	3–4	P, N	C/A/M	UPL	7	9/11–10/8	8/8–9/6	
Helianthus occidentalis, western sunflower (2 ssp.)	3–5	P, S, N	C/A	FACU–	7	9/17–10/30	8/18	
**Helianthus pauciflorus*, rigid/stiff sunflower (2 ssp.)	3–4	P	C/A	UPL	6	10/5–10/18	8/21	
Helianthus petiolaris, prairie sunflower (2 ssp.)	4–5	P, N		UPL	*			
Helianthus strumosus, pale-leaved sunflower	3–5	P, S		UPL	3			
**Helianthus tuberosus*, Jerusalem artichoke	2–4	P, S	C/A	FAC	3	10/21	9/17	
**Heliopsis helianthoides*, ox-eye (3 var.)	2–4	P, S	C/A/M	UPL	4	8/10–10/6	6/18	
Heliotropium tenellum, glade heliotrope	4–5	P		UPL	10			
**Heracleum maximum*, cow-parsnip	2–3	P, S, W		FACW	6	6/20–7/20	6/10	
**Hesperostipa spartea*, porcupine grass	3–5	P, S, N	C:G/A:G/M	UPL	6	6/14–6/20	5/25	
Heterotheca camporum, golden aster (2 var.)	4–5	P, N	C	UPL	5		8/23	
**Heuchera richardsonii*, prairie alumroot	2–5	P, S	A:L:D	FAC–	7	7/2–7/17	5/16	DF*
Hibiscus laevis, halberd-leaved rose mallow	1–2	P	C	OBL	4		8/10	
Hibiscus lasiocarpos, hairy rose mallow	1–2	P, S	A	FACW+	5			
Hibiscus moscheutos, swamp rose mallow	1–2	P	L	OBL	6		8/21	
Hieracium gronovii, hairy hawkweed	4–5	P, S		UPL	5		8/15	
Hieracium longipilum, hairy hawkweed	4–5	P	C	UPL	6	8/1–	7/25	

(continued)

(continued)

	H_2O	Eco	Seed	Wet	STAT/CC	Seed-time	FLR-time	PREDTN
Hieracium scabrum, rough hawkweed (3 var.)	3–4	P, S		UPL	5		8/17	
Hordeum jubatum, squirrel-tail grass (2 ssp.)	3–4	P		FAC+	*		7/15	
Hordeum pusillum, little wild barley	3–4	P		FAC	0		6/9	
Houstonia caerulea, bluets	3–4	P, S	A	FAC	7		5/30	
Houstonia longifolia, long-leaved houstonia	4–5	P, S, N	A:D	UPL	7			
Houstonia purpurea, broad-leaved bluets (3 var.)	3–5	P, S, N		UPL	10			
Houstonia pusilla, tiny bluets	4–5	P, S		FACU–	3			
Hymenopappus scabiosaeus, old plainsman	4–5	P, S, N		UPL	9			
**Hypericum ascyron*, great St. John's-wort	2–3	P, S	C:D/M:D	FAC+	8	9/14–10/15	7/8	
Hypericum canadense, Canadian St. John's-wort	2–3	P, N		FACW	8		8/3	
Hypericum drummondii, nits-and-lice	3–4	P, S		FACU	6			
Hypericum gentianoides, pineweed	4–5	P, S, N		FACU	6		8/10	
Hypericum gymnanthum, small-flowered St. John's-wort	2–3	P, S		OBL	9			
Hypericum kalmianum, Kalm's St. John's-wort	2–3	P, N		FACW–	10		7/19	
Hypericum majus, small-flowered St. John's-wort	2–3	P		FACW	7		8/8	
Hypericum mutilum, dwarf St. John's-wort	2–3	P, S		FACW	5		9/15	
Hypericum prolificum, shrubby St. John's-wort	2–4	P, S, N		FACU	6			
**Hypericum punctatum*, spotted St. John's-wort	3–4	P, S	C:D/M:D	FAC+	3	9/16–10/25	8/10	
Hypericum sphaerocarpum, round-fruited St. John's wort	3–4	P, S		FACU	5		7/14	
Hypoxis hirsuta, yellow star-grass	2–4	P, S, N	C:D	FAC	6	6/5–7/10	5/8	
Iliamna rivularis v. *rivularis (I. remota)*, Kankakee mallow	2–4	P, N	C	UPL	E/10	8/11–9/29	6/16	
Ionactis linariifolius, stiff aster	4–5	P, S	A	UPL	9		9/15	
Ipomoea lacunosa, small morning glory	2–4	P, S		FACW	1		9/12	
Iris brevicaulis, blue water iris	1	P		OBL	9			
Iris virginica, v. *shrevei*, southern blue flag	1–2	P, S	M/C120	OBL	5	8/8–8/21	6/16	
Isolepis carinata, keeled bulrush	2	P		FACW+	8			
Juncus acuminatus, knotty-leaved rush	1–2	P		OBL	4			

Juncus alpinoarticulatus ssp. *nodulosus*, Richardson's rush (2 ssp.)	2	P		OBL	8		
Juncus articulatus, jointed rush	2	P		OBL	9		
Juncus biflorus, two-flowered rush	1–2	P		FACW	5		
Juncus brachycarpus, short-fruited rush	1–2	P	C:D	FACW	5		
Juncus brachycephalus, short-headed rush	1–2	P		OBL	7		
Juncus bufonius, toad rush (3 var.)	1–2	P		FACW+	2		
Juncus canadensis, Canadian rush	1–2	P		OBL	6		
Juncus dudleyi, Dudley rush	1–2	P	C:D	FAC	4		
Juncus effusus, common rush (9 var.)	1–2	P	O:D	OBL	4		
Juncus greenei, Greene's rush	1–2	P, N		FAC	7		
Juncus interior, inland rush (3 var.)	1–2	P	C:D	FAC+	3		
Juncus marginatus, grass-leaved rush	1–2	P, N		FACW	5		
Juncus nodosus, joint rush (2 var.)	1–2	P		OBL	6		
Juncus scirpoides, round-headed rush	1–2	P, N		FACW+	9		
Juncus secundus, secund rush	2–4	P		FAC–	6		
**Juncus tenuis*, path rush	2–5	P, S	O:D	FAC	0	7/13	6/13
Juncus torreyi, Torrey rush	1–2	P	C:D	FACW	3		
Juncus vaseyi, Vasey's rush	2–3	P, S		FACW	10		
Koeleria macrantha, June grass	4–5	P, S, N	A:G:D/M:D	UPL	7	6/26–7/7	6/8
Krigia biflora, false dandelion (2 var.)	2–3	P, S		FACU	5	6/16–	5/22
Krigia caespitosa, opposite-leaved dwarf dandelion	2–3	P		FAC–	1		
Krigia dandelion, dwarf dandelion	2–3	P, S, N		FACU	6		
Krigia virginica, dwarf dandelion	4–5	P, N		UPL	4		5/20
Kuhnia see *Brickellia*							
Lactuca ludoviciana, prairie/western lettuce	3–4	P		UPL	E/10		
Lactuca tatarica, large-flowered blue lettuce (2 var.)	2–3	P, S		–	–		7/10
Lathyrus palustris, marsh vetchling	1–2	P, S		OBL	7		7/3

(continued)

(continued)

	H_2O	Eco	Seed	Wet	STAT/CC	Seed-time	FLR-time	PREDTN
Lathyrus venosus, veiny pea	3–4	P, S, N	C10:H:I	FAC	9	8/1–	6/15	
Lechea intermedia, savanna pinweed (3 var.)	4–5	P, S		UPL	10			
Lechea mucronata, hairy pinweed	4–5	P, S, N		UPL	7			
Lechea stricta, pinweed	4–5	P, N		UPL	8			
Lechea tenufolia, narrow-leaved pinweed	4–5	P, N		UPL	6			
Lespedeza, several hybrids, × *longifolia (capitata × hirta)*, × *manniana (capitata × violacea)*, × *nuttallii (hirta × violacea)*, × *simulata (capitata × virginica)*								
**Lespedeza capitata*, bush lespedeza	3–5	P, S	C10:J/M	FACU	4	9/1–10/17	8/14	DP*
Lespedeza frutescens, shrubby bush-clover	4–5	S		UPL	6		8/31	
Lespedeza hirta, hairy bush-clover (2 spp.)	4–5	P, S, N		UPL	6		8/31	
Lespedeza leptostachya, prairie bush-clover	4–5	P, N	C10:J	UPL	E, FT/10		10/6	
Lespedeza procumbens, trailing bush-clover	4–5	P, S, N		UPL	5			
Lespedeza repens, creeping bush-clover	4–5	P, S, N		UPL	6			
Lespedeza stuevei, Stuve's bush-clover	4–5	P, S, N		UPL	6			
Lespedeza virginica, Virginia lespedeza/slender bush clover	4–5	P, S	C10:J	UPL	5		8/22	
Lesquerella ludoviciana, silvery bladderpod	4–5	P, N		UPL	10			
Leucospora multifida, conobea	1–2	P, N		FACW+	3		8/20	
Liatris, hybrids, × *steelei (spicata × aspera)*, × *ridgewayi (pycnostachya × squarrosa* or *aspera)*								
**Liatris aspera*, rough blazing star (3 var.)	3–5	P, S, N	C/M	UPL	7	9/12–10/23	8/24	MC
Liatris cylindracea, cylindrical blazing star	4–5	P, S, N	C/M	UPL	8	9/26–10/8	8/15	
Liatris punctata, dotted button-snakeroot (3 var.)	4–5	P, N	C	UPL	*		7/10	
**Liatris pycnostachya*, gayfeather (2 var.)	1–3	P	C/M	FAC–	6	8/27–10/25	8/8	MC
Liatris scariosa, savanna blazing star (3 var.)	3–5	P, S, N	C	UPL	T/7			
**Liatris spicata*, marsh blazing star (2 var.)	1–3	P, S	C90	FAC	7	10/5–10/18	8/4	
Liatris squarrosa, scaly blazing star (5 var.)	4–5	P, S	C/M	UPL	7	9/1–10/6	8/20	
Lilium michiganense, Michigan lily	1–3	P, S	L/O:C90	FAC+	6	8/19–9/2	7/4	
Lilium philadelphicum, western/wood lily (2 var.)	3–4	P, S, N	C	FAC–	10		6/29	
Linum medium, small yellow flax (2 var.)	2–4	P, S, N		FACU	7		7/25	
Linum striatum, stiff yellow flax	2	P, S, N		FACW–	8			

Linum sulcatum, grooved yellow flax (2 var.)	4–5	P, N	A/C	UPL	8		8/27
Liparis loeselii, green twayblade	2–3	P, S		FACW+	8		
Lipocarpha micrantha, small-flowered rush	1–2	P, N		OBL	7		
**Lithospermum canescens*, hoary puccoon	3–5	P, S, N		UPL	6	8/1–	5/13
Lithospermum caroliniense, pucoon (2 var.)	4–5	P, S, N		UPL	7	8/1–	5/22
Lithospermum incisum, fringed puccoon	4–5	P, S, N	A:D	UPL	8		5/23
**Lobelia cardinalis*, cardinal flower	1–2	P, S	O:D/C90:D	OBL	6	9/1–10/20	9/7
Lobelia kalmii, Kalm's lobelia	1–2	P, S		OBL	10		8/20
Lobelia puberula, downy lobelia (4 var.)	2–3	P, S, N		OBL	8		
**Lobelia siphilitica*, blue cardinal-flower (2 var.)	1–3	P, S	A:D/M:D	FACW+	4	9/1–11/3	8/15
**Lobelia spicata*, spiked lobelia (5 var.)	2–4	P, S	C:D/M:D	FAC	4	7/20–10/27	6/10
Ludwigia alternifolia, seedbox	1–2	P		OBL	5		7/31
Ludwigia palustris, marsh purslane	1–2	P		OBL	4		8/10
Ludwigia polycarpa, false loosestrife	1–2	P, S		OBL	5		8/22
Lupinus perennis, wild lupine (2 ssp., 2 var.)	4–5	P, S, N	A:H:G/M	UPL	8	6/5–6/19	5/19
Luzula multiflora, wood rush (4 var.)	3–4	P, S		FACU	5		
Lycopus americanus, cut-leaved water horehound	1–2	P, S	A:D	OBL	3		8/10
Lycopus uniflorus, northern bugle weed	1–2	P		OBL	7		
**Lysimachia ciliata*, fringed loosestrife	2–3	P, S	C/A	FACW	4	9/25–10/25	7/10
Lysimachia hybrida, loosestrife	1–3	P, S	A	OBL	7		
Lysimachia lanceolata, lance-leaved loosestrife	3–4	P, S		FAC	6		7/20
Lysimachia × producta, loosestrife (*quadrifolia × terrestris*)	1–2	P, S		–	–		
**Lysimachia quadriflora*, prairie loosestrife	1–2	P, S	C:D	OBL	8	9/25–10/15	6/25
Lysimachia quadrifolia, four-leaved loosestrife	3–4	P, S		UPL	9		6/4
Lysimachia terrestris, swamp loosestrife (2 var.)	1–2	P		OBL	8		7/10
Lysimachia thyrsiflora, tufted loosestrife	1–2	P	C:D	OBL	7		6/5
**Lythrum alatum*, winged loosestrife (2 var.)	1–2	P	C:D/M:D	OBL	5	8/8–10/16	6/25
**Maianthemum racemosum* ssp. *racemosum*, starry solomon's plume (2 ssp.)	2–5	P, S, W	L/F	FACU	4		5/10

(continued)

(continued)

	H_2O	Eco	Seed	Wet	STAT/CC	Seed-time	FLR-time	PREDTN
**Maianthemum stellatum*, starry false Solomon's seal (2 var,)	2–4	P, S, W		FAC–	5			
Manfreda virginica, American aloe	4–5	P, S		UPL	8			
Melanthium see *Veratrum*								
Melica nitens, 3-flowered melic grass	3–4	P, S		UPL	7		6/5	
Mentha arvensis, wild mint	2–3	P		FACW	4			
Mentzelia oligosperma, stickleaf	4–5	P		UPL	10			
Mimosa microphylla, sensitive-brier	4–5	P, S, N	C10:H/M	UPL	7	8/14–10/13	7/10	
Minuartia michauxii, stiff sandwort	4–5	P, N	A	UPL	–		6/20	
Minuartia patula, slender sandwort (2 form)	4–5	P, S, N		UPL	T/–		5/30	
Mirabilis hirsuta, hairy umbrellawort	4–5	P		UPL	5			
**Mirabilis nyctaginea*, wild four-o'clock	3–4	P	A/M	UPL	–	8/24–9/12	6/9	
Monarda bradburiana, monarda	3–4	P, S	C	UPL	5			
**Monarda fistulosa*, wild bergamot (2 ssp., 5 var.)	2–3	P, S	M:D/O:D/A:D	FACU	4	9/11–10/8	7/14	
Monarda punctata, spotted bee balm (2 ssp., 7 var.)	4–5	P, S, N	O:D/M:D:Q	UPL	5	9/5–10/16	8/6	
Muhlenbergia capillaris, hair grass	3–4	P, S, N		FACU	8			
Muhlenbergia cuspidata, prairie satin grass	3–5	P	A:D	UPL	10			
Muhlenbergia frondosa, common satin grass	2–3	P, S		FACW	3			
Muhlenbergia glabriflora, smooth satin grass	3–5	P, S, N		UPL	7			
Muhlenbergia glomerata, marsh wild timothy	1–2	P	A:D	OBL	10			
Napaea dioica, glade mallow	2–3	P, S	C	FACW–	4	8/18–9/14	7/1	DPF*
Nothocalais cuspidata, prairie dandelion	4–5	P, N	C30:G	UPL	E/9	5/15–5/30	5/1	
Nothoscordum bivalve, yellow false garlic	3–5	P, S, N		UPL	5			
**Oenothera biennis*, evening primrose	2–4	P, S	A:D/C90:D	FACU	1	earlier–11/21	7/19	
Oenothera fruticosa, shrubby sun drops (2 ssp.)	3–4	P, S		FACU+	9			
Oenothera laciniata, ragged evening primrose (2 var.)	3–5	P, S, N		FACU	2		6/30	
Oenothera linifolia, thread-leaved sundrops	4–5	P, S, N		UPL	8			
Oenothera perennis, small sundrops (2 var.)	2–4	P		FAC	E/8		7/3	
Oenothera pilosella, prairie sundrops (2 ssp.)	2–4	P, S		FAC–	6		6/20	

Oenothera rhombipetala, sand primrose	4–5	P, S, N	D	FACU	5		7/17
Oenothera speciosa, showy evening primrose (west)	4–5	P, S, N	A	UPL	*		7/1
Oligoneuron album, stiff aster	4–5	P, N	A	UPL	9	9/16–	8/10
Oligoneuron ohioense, Ohio goldenrod	1–2	P	C30:D	OBL	10		7/20
Oligoneuron riddellii, Riddell's goldenrod	1–3	P	C:D	OBL	7		10/1
**Oligoneuron rigidum*, rigid goldenrod (3 var.)	3–5	P, S	C/M	FACU–	4	10/8–10/23	9/7
Onosmodium bejariense v. *bejariense*, marbleseed	3–5	P, S, N	M:F/M:Q	UPL	10		
Onosmodium bejariense v. *hispidissimum*, shaggy false gromwell	3–5	P, S, N	M:F/M:Q	UPL	5		6/17
Onosmodium bejariense v. *occidentale*, western false gromwell	3–5	P, S, N	M:F/M:Q	UPL	8	9/5–10/18	7/4
Opuntia fragilis, little prickly pear	4–5	P, S		UPL	9		
Opuntia humifusa, prickly pear cactus	5	P, S, N	A	UPL	5	9/3	6/18
**Orbexilum onobrychis*, French grass	3–4	P, S	C10:H	UPL	6	8/12–9/25	7/3
Orbexilum pedunculatum, Sampson's snakeroot (2 var.)	4–5	P, S	C10:H	UPL	6		
Orobanche fasciculata, clustered broomrape	4–5	P, N	K	UPL	E/10		6/12
Orobanche ludoviciana, broomrape (2 ssp.)	4–5	P, N	K	UPL	E/10		
**Oxalis violacea*, purple oxalis	3–4	P, S, W	M/C	UPL	5	6/5–6/20	5/2
Oxypolis rigidior, cowbane	1–2	P, S		OBL	7		8/25
Packera aurea, golden ragwort (5 var., 2 forms)	2	P, S	A:D	FACW	4		5/20
**Packera paupercula*, balsam groundsel	1–4	P, S, N	C	FAC+	3	6/24–7/1	6/5
Packera plattensis, prairie groundsel	3–4	P	C:D	FACU–	6		5/14
Panicum also see *Dichanthelium*							
Panicum anceps, panic grass	1–2	P, S		FACW	3		
Panicum flexile, slender panic grass	1–2	P, S, N		FACW+	7		
Panicum gattingeri, panic grass	2–4	P		FAC	5		
Panicum philadelphicum	4–5	P, N		UPL	5		
Panicum rigidulum, munro grass (4 var.)	1–2	P, S, N		FACW	6		

(continued)

(continued)

	H_2O	Eco	Seed	Wet	STAT/CC	Seed-time	FLR-time	PREDTN
**Panicum virgatum*, prairie switchgrass (2 var.)	2–5	P, S	A/M	FAC+	4	8/28–10/22		
Parnassia glauca, grass-of-parnassus	1–2	P	C90:D	OBL	9		9/5	
Paronychia canadensis, tall forked chickweed	4–5	P, S, N		UPL	5			
**Parthenium integrifolium*, American feverfew/wild quinine (2 var.)	2–4	P	C/M	UPL	8	8/21–10/17	6/25	MR
Paspalum fluitans, swamp bead grass	1–2	P		OBL	5			
Paspalum laeve, smooth lens grass (2 var.)	1–2	P		UPL	2			
Paspalum pubiflorum, hairy-flowered bead grass	1–2	P		FACW	3			
Paspalum setaceum, bead grass	3–5	P, S, N	A	UPL	3			
Pedicularis canadensis, wood betony (2 ssp.)	2–5	P, S	C30:K	FACU+	7	6/1–	5/5	
**Penstemon digitalis*, foxglove beard-tongue	2–4	P, S	N:D/O:D/M:D	FAC–	4	9/4–10/3	6/10	
Penstemon grandiflorus, large-flowered beard-tongue	4–5	P, N	N:G/O:G/M	UPL	E/8	8/6–8/14	6/5	
Penstemon hirsutus, hairy beard-tongue	4–5	P, S, N	C90:D	UPL	8		6/6	
**Penstemon pallidus*, pale beard-tongue	4–5	P, S, N	C30:G:D	UPL	6		6/19	
Penstemon tubiflorus, funnel-form beard-tongue	3–5	P, S	C30:D	UPL	5		7/20	
Petalostemum see *Dalea*								
Phemeranthus calycinus, fame flower	4–5	P, N	C90	UPL	E/10			
Phemeranthus parviflorus, prairie fame flower	4–5	P, N		UPL	10			
Phemeranthus rugospermus, fame flower	5	P, N	C:D	UPL	9		7/15	
Phlox bifida, sand or ten-point phlox (2 ssp.)	4–5	P, S, N	C	UPL	7		5/22	
Phlox glaberrima v. *interior*, smooth phlox (3 ssp.)	2–3	P, S	N:C90/C90:Q	FACW	6	9/16–	7/20	
**Phlox maculata*, wild sweet William (2 ssp.)	2	P, S	C	OBL	10		9/10	
**Phlox pilosa*, downy phlox (9 ssp.)	3–5	P, S, N	C90:Q	FAC–	7	7/16–	5/19	
Phragmites australis, common reed (worldwide)	1–2	P		FACW+	1			
**Physalis heterophylla*, clammy ground cherry (2 var.)	3–4	P, S		UPL	2		6/9	
Physalis pubescens, tall hairy ground-cherry (2 var.)	3–5	P, N		UPL	4			
Physalis pumila, dwarf ground- cherry	4–5	P, S, N		UPL	5			
Physalis virginiana, Virginia ground-cherry (4 var.)	3–4	P, S, N	R:D	UPL	3	9/16–	6/15	

Physostegia parviflora, western obedient plant	1–3	P, S	C	FACW–	–	8/1–9/20	7/1	
Physostegia speciosa, showy obedient plant	1–3	P, S		FACW	7			
**Physostegia virginiana*, obedient plant (2 ssp.)	1–3	P, S	A	FACW	6	10/8–11/18	8/8	
Plantago aristata, large-bracted plantain (annual)	4–5	P		UPL	1		8/14	
Plantago patagonica, wooly plantain (annual)	5	P, N	C30	UPL	–		6/19	
Plantago pusilla, small plantain (annual)	3–5	P, S, N		FACU	3			
**Plantago virginica*, dwarf plantain (annual)	3–5	P, S	A:D	FACU–	3		6/2	
Platanthera ciliaris, orange fringed orchid	2–3	P, S		FACW	10		7/30	
Platanthera clavellata, green orchid	1–2	P, S		OBL	10		7/28	
Platanthera flava, tubercled orchid (2 var.)	2–3	P, S		FACW	10		7/1	
Platanthera lacera, green fringed orchid	2–3	P, S		FACW	9		7/10	
Platanthera leucophaea, prairie white fringed orchid	1–2	P		FACW+	E, FT/10		6/19–7/7	
Platanthera peramoena, purple fringeless orchid	1–2	P, S		FACW	7			
Platanthera psycodes, purple fringed orchid	2–3	P, S		FACW	10			
Poa chapmaniana, annual bluegrass	2–4	P		FACU	1		5/20	
Poa palustris, marsh bluegrass	1–2	P, S		FACW+	7		7/27	
Pogonia ophioglossoides, rose pogonia (2 var., 2 form)	1–2	P, S		OBL	E/10		6/26	
Polanisia jamesii, James' clammyweed	4–5	P, N		UPL	E/5			
Polemonium reptans, Jacob's-ladder (2 var.)	2–4	P, S, W	E:Q/M	FAC	5	5/31–6/15	4/21	DPF*
Polygala incarnata, pink milkwort (annual)	4–5	P, S, N		FACU–	E/10		8/30	
Polygala polygama v. *obtusata*, bitter milkwort	3–4	P		FACU–	7			
**Polygala sanguinea*, field milkwort (annual)	2–4	P, S	L	FACU	5	8/1–8/15	8/1	
Polygala senega, Seneca snakeroot	3–5	P, S, N	C90	FACU	7		5/25	
Polygala verticillata, whorled milkwort (annual, 2 var.)	2–4	P, S, N		UPL	5-9		8/4	
Polygonatum biflorum, small Solomon's seal (5 var.)	2–4	S, W	R:M	FACU	7			
**Polygonatum biflorum* v. *commutatum*, great Solomon's seal	2–4	S, W	L:F/M:Q	FACU	4	8/16	6/12	
Polygonella articulata, jointweed	4–5	P, N		UPL	9			

(continued)

(continued)

	H_2O	Eco	Seed	Wet	STAT/CC	Seed-time	FLR-time	PREDTN
Polygonum amphibium, water knotweed (2 var.)	1–2	P		OBL	3		7/9	
Polygonum arifolium, halberd-leaved tear-thumb	1–2	P		OBL	10			
Polygonum careyi, Carey's heartsease	1–2	P, S		FACW+	E/10		7/31	
Polygonum hydropiperoides, mild water pepper	1–2	P		OBL	4		8/18	
Polygonum pensylvanicum, pink smartweed	2–3	P		FAC	2			
Polygonum scandens, climbing false buckwheat (3 var.)	2–4	P, S, N		FAC	4			
Polygonum tenue, slender knotweed (annual.)	4–5	P		UPL	5		8/30	
Polytaenia nuttallii, prairie parsley	3–5	P, S	M/C120	UPL	8	7/30–8/14	5/21	
**Potentilla arguta*, prairie cinquefoil (2 ssp.)	2–5	P, S	C:G:D/M:D	FACU–	10	8/8–9/5	7/4	DP*
**Potentilla canadensis*, common cinquefoil (2 var.)	2–4	P, S, W		–	–			
Potentilla pensylvanica, gray cinquefoil	4–5	P		UPL	10			
Potentilla rivalis, cinquefoil (annual)	1–2	P		OBL	E/10			
**Potentilla simplex*, common cinquefoil	2–4	P, S		FACU–	3		5/22	
**Prenanthes aspera*, rough white lettuce	3–5	P	C	UPL	8		8/28	
Prenanthes racemosa, rattlesnake root (2 var.)	2–4	P	C120/M	FACW	8		8/20	
Pseudognaphalium obtusifolium, sweet everlasting (3 spp.)	3–5	P, S	C:D	UPL	2		9/8	
Psoralea see *Orbexilum* and *Psoralidium*								
Psoralidium tenuiflorum, scurf-pea	4–5	P, S, N	C10:H	UPL	8	7/24–8/25	6/27	DP*
Pulsatilla patens, pasque flower (2 ssp.)	4–5	P, S, N	C	UPL	9	5/15–	4/15	
Pycnanthemum muticum, broad-leaved mtn. mint	2–4	P, S, W, N		FAC	10			
**Pycnanthemum tenuifolium*, slender mountain mint	2–4	P, S	A:D/M:D	FAC	4	8/24–10/30	7/5	
**Pycnanthemum verticillatum* v. *pilosum*, hairy mountain mint	2–4	P, S	A:D/M:D	UPL	6	9/18–10/31	7/14	
**Pycnanthemum virginianum*, common mountain mint	1–3	P, S	A:D/M:D:Q	FACW+	5	9/5–10/31	7/31	
Pyrrhopappus carolinianus, false dandelion (annual/biennial)	3–5	P, N		UPL	1			
Ranunculus fascicularis, early buttercup (2 var.)	3–4	P, S		FACU	5			
Ranunculus laxicaulis, water plantain spearwort	1–2	P, S		OBL	6			

Ranunculus rhomboideus, prairie buttercup	4–5	P, N	L:M/C	UPL	T/10		
**Ratibida pinnata*, drooping coneflower	2–4	P	C/A/M	UPL	4	8/14–9/25	7/5
Rhexia mariana, Maryland meadow beauty (3 var.)	1–2	P, S, N	A:D	OBL	10		
Rhexia virginica, meadow beauty	1–2	P, S, N	C90:D/O:D	OBL	10		8/7
Rhynchospora alba, white beak rush	1–2	P, N		OBL	10		
Rhynchospora capillacea, hair beak rush	1–2	P, N		OBL	10		
Rhynchospora capitellata, brown beak rush	1–2	P		OBL	10		
Rhynchospora globularis, grass beak rush (2 var.)	2	P, N		FACW	10		
**Rosa arkansana*, prairie/sunshine rose (2 var.)	2–5	P, S	C:H:F	UPL	5	8/12–9/14	5/18
**Rosa carolina*, pasture rose (2 var.)	2–5	P, S	C90:Q	FACU–	4	8/25–10/23	6/9
Rosa palustris, swampy rose	2	P, S	C90:Q	OBL	7		6/25
**Rosa setigera*, prairie/climbing rose (2 var.)	2–4	P, S	C	FACU+	5		7/10
Rosa virginiana, Virginia rose (2 var.)	2–4	P, S	E	FAC	–	10/22	6/24
Rudbeckia fulgida, orange coneflower (5 var.)	2–4	P, S	C	–	7	9/3–10/15	7/26
Rudbeckia fulgida v. *sullivantii*, Sullivant's orange coneflower	2–3	P, S	S/M:Q	OBL	6		
**Rudbeckia hirta*, black-eyed susan (biennial, 4 var.)	2–5	P, S	C/A/M	FACU	2	8/6–8/16	7/4
Rudbeckia missouriensis, Missouri orange coneflower	2–4	P, S, N		FACU–	E/10		
**Rudbeckia subtomentosa*, fragrant coneflower	2–3	P, S	C/A/M	FACW	5	9/18–11/8	8/8
**Rudbeckia triloba* v. *triloba*, branched coneflower	2–3	S	A	FAC–	3		8/25
**Ruellia humilis*, hairy ruellia (2 var.)	3–5	P, S	C100/M	FACU–	3	10/1–10/17	7/5
Rumex altissimus, pale dock	2–3	P, S	A	FACU–	2		6/18
Rumex hastatulus, Engelmann's sorrel/sour dock	2–3	P, N		FACU	E/4		
Rumex orbiculatus, great water dock	1–2	P	C	OBL	7		7/1
Rumex salicifolius v. *mexicanus*, Mexican dock	2	P, N		FAC+	1		6/28
Rumex verticillatus, swamp dock	1–2	P		OBL	5		9/18
Sabatia angularis, rose gentian (annual)	2–3	P, S		FAC+	3		7/31
Sabatia campestris, prairie rose gentian (annual)	3–4	P, N		FACU	E/8		

(continued)

(continued)

	H_2O	Eco	Seed	Wet	STAT/CC	Seed-time	FLR-time	PREDTN
Salix humilis, prairie willow (2 var.)	3–4	P, N		FACU	5	5/20–5/25	4/15	DP*
Salvia azurea v. *grandiflora*, blue sage	4–5	P, S	A	UPL	T/9		9/13	
Salvia lyrata, cancer weed	3–4	P, S, N		FACW–	4			
Sanguisorba canadensis, American burnet	2–3	P	C	FACW+	E/10		9/2	
Sanicula canadensis, Canadian black snakeroot	3–4	S		FACU+	4			
Sanicula marilandica, black snakeroot	4–5	P, S		UPL	6			
Saxifraga pensylvanica, swamp saxifrage	1–2	P, S	C:D	OBL	10	6/16–	5/30	
Schedonnardus paniculatus, tumblegrass	3–5	P		UPL	5			
**Schizachyrium/Andropogon scoparium*, little bluestem grass	3–5	P, S	A/M	FACU–	5	9/27–11/11		
Schoenoplectus acutus, hard-stemmed bulrush (2 var.)	1–2	P		OBL	6			
Schoenoplectus americanus, chairmaker's rush	1–2	P		OBL	3			
Schoenoplectus fluviatilis, river bulrush	1–2	P	C90/M	OBL	3			
Schoenoplectus hallii, Hall's bulrush	1–2	P, N		OBL	10			
Schoenoplectus heterochaetus, slender bulrush	1–2	P		OBL	7			
Schoenoplectus smithii, Smith's tufted bulrush	1–2	P		OBL	E/10			
Schoenoplectus tabernaemontani, great bulrush	1–2	P	C:D	OBL	–			
Shrankia see *Mimosa*								
Scirpus also see *Schoenoplectus* and *Isolepis*								
Scirpus atrovirens, dark green rush	1–2	P	C:D/M:D	OBL	4			
**Scirpus cyperinus*, wool grass	1–2	P, S	C:D/M:D	OBL	5	7/1–7/10		
Scirpus georgianus, bristleless dark green rush	1–2	P		OBL	4			
**Scirpus pendulus*, red bulrush	1–2	P, S		OBL	3	8/6	5/29	
Scleria pauciflora, few-flowered nut rush	2–4	P, S		FACU	10			
Scleria triglomerata, tall nut grass	2–4	P, S		FAC	9			
Scleria verticillata, low nut grass	1–2	P		OBL	10			
Scuttellaria galericulata, marsh skullcap	1–2	P, S.N		OBL	6		7/15	
Scutellaria lateriflora, mad-dog skullcap	1–2	P, S	A	OBL	4		7/27	

**Scutellaria parvula*, small skullcap (3 var.)	3–5	P, S, N		FACU	6			
Senecio see *Packera*								
Senna hebecarpa, wild senna (2 var.)	2–3	P, S	C10:H	FACW	9		8/2	
Senna marilandica, Maryland senna	2–3	P, S	C10:H/M	FACW	4	10/1–11/16	8/3	
Setaria parviflora, perennial foxtail	2–4	P		FAC	6			
**Silene antirrhina*, sleepy catchfly	3–5	P, S		UPL	1		6/17	
Silene regia, royal catchfly	3–4	P, S	C/M	UPL	E/9	8/8–9/5	7/19	
**Silene stellata*, starry campion	2–4	P, S	C/M	UPL	6	7/25–8/20	6/27	
**Silphium integrifolium*, rosinweed (5 var.)	2–4	P	C/M	UPL	5	8/21–9/18	7/19	
**Silphium laciniatum*, compass plant (2 var.)	2–4	P	C/M	FACU–	5	9/18–10/22	8/4	MR*
**Silphium perfoliatum*, cup plant (2 var.)	2–3	P, S	C/M	FACW–	4	8/14–9/15	7/24	
**Silphium terebinthinaceum*, prairie dock (2 var.)	2–4	P	C/M	FAC–	4	9/5–10/23	8/8	DP*
Silphium terebinthinaceum × *laciniatum*, hybrid	2–4	P		FACU	–			
**Sisyrinchium albidum*, common blue-eyed grass	3–5	P, S, N	M:D/C:G:D	FACU	4	6/18–7/8	5/11	
**Sisyrinchium angustifolium*, narrow-leaved blue-eyed grass	2–4	P, S	D/M:D	FACW–	5	7/1–7/17	6/10	
Sisyrinchium atlanticum, eastern blue-eyed grass	2–3	P, S, N		FACW	E/10		6/21	
Sisyrinchium campestre, prairie blue-eyed grass (2 forms)	3–5	P, N	P:Q:D	UPL	6			
Sisyrinchium montanum, mountain blue-eyed grass (2 var.)	2–4	P, S		FAC+	E/9		6/11	
Sisyrinchium mucronatum, blue-eyed grass	2–4	P, S		FACW–	9			
Sium suave, water parsnip/hemlock	1–2	P, S		OBL	5		8/15	
Smilacina see *Maianthemum*								
Solidago also see *Oligoneuron* and *Euthamia*								
Solidago bicolor, silverrod (2 var.)	4–5	P, S		UPL	7			
**Solidago canadensis*, tall/Canada goldenrod (5 var.)	2–4	P, S	N/O	FACU	1	10/24–11/14	9/17	
Solidago hispida v. *hispida*, white goldenrod	3–5	P, S		UPL	7			
**Solidago juncea*, early goldenrod	3–5	P, S	C:D	UPL	4	9/17	8/18	
Solidago missouriensis v. *fasciculata*, Missouri goldenrod	4–5	P, S, N		UPL	4			
**Solidago nemoralis*, old field goldenrod (2 var.)	4–5	P, S	C:D	UPL	3	10/31	8/30	

(continued)

(continued)

	H_2O	Eco	Seed	Wet	STAT/CC	Seed-time	FLR-time	PREDTN
Solidago patula, rough-leaved goldenrod (2 var.)	1–2	P	C:D:G	OBL	9		9/5	
**Solidago speciosa*, showy goldenrod (4 var.)	3–5	P, S	C:D/M:D	UPL	7	10/16–11/29	10/3	MR(?)
Solidago uliginosa, bog goldenrod (4 var.)	1–2	P, S	C:D	OBL	10		9/15	
**Sorghastrum nutans*, Indian grass	3–5	P, S	A/M	FACU+	4	9/26–10/30	9/7	
Sparganium americanum, American bur reed	1	P		OBL	10		6/23	
Sparganium androcladum, branched bur reed	1	P		OBL	10			
Sparganium eurycarpum, common bur reed	1	P	F	OBL	5		7/10	
**Spartina pectinata*, freshwater cord grass	1–3	P	A:Q	FACW+	4	10/18–10/25	8/10	
Sphenopholis intermedia, slender wedge grass	2–4	P		FAC	5			
Sphenopholis obtusata, prairie wedge grass	2–4	P, S		FAC	5		6/16	
Spiraea alba, meadow-sweet (2 var.)	1–2	P, S	O/C90	FACW+	6		7/14	
Spiranthes cernua, nodding ladies tresses	2–3	P, S		FACW–	4	10/10–10–25	9/10	
Spiranthes lacera, slender ladies' tresses	2–4	P, S		FAC+	7		8/29	
Spiranthes lucida, early ladies' tresses	1–3	P, S		FACW+	E/10		6/15	
Spiranthes magnicamporum, great plains ladies'tresses	2–3	P		FACW	6			
Spiranthes romanzoffiana, hooded ladies' tresses	1–2	P, S		FACW+	E/10			
Spiranthes tuberosa, little ladies' tresses	4–5	P, S		UPL	9	9/20	8/24	
Spiranthes vernalis, spring ladies' tresses	2–3	P, S		FAC	E/7			
Sporobolus clandestinus, dropseed	3–5	P, S, N		UPL	6			
**Sporobolus compositus*, dropseed (3 var.)	3–5	P, N	A/M	UPL	3	9/25–10/17		
Sporobolus cryptandrus, sand dropseed	3–5	P, N	A	FACU–	4			
**Sporobolus heterolepis*, northern dropseed	2–4	P, S	A/M	FACU–	9	8/14–10/8	7/4	
Sporobolus neglectus, dropseed	4–5	P, N		UPL	1			
Sporobolus vaginiflorus, northern rush/poverty grass (2 var.)	4–5	P, N		UPL	0			
Stachys aspera, rough hedge nettle	1–2	P, N		FACW+	9		7/14	
Stachys palustris, woundwort	1–2	P		OBL	5		7/18	
Stachys tenuifolia, rough hedge nettle (4 var.)	1–2	P, S		OBL	5			
Stenaria nigricans, narrow-leaved bluets	4–5	P, S, N		UPL	7			

Stipa see *Hesperostipa*								
Strophostyles leiosperma, wild bean (annual)	3–5	P, S, N		UPL	4		8/20	
Strophostyles umbellata, clustered wild bean (2 var.)	3–4	P, S, N		FACU	5			
Stylisma pickeringiii, Pickering's dawnflower (2 var.)	4–5	P, N		UPL	T/9			
Stylosanthes biflora, pencil flower	4–5	P, S, W		UPL	5			
Symphoricarpos occidentalis, wolfberry	4–5	P		UPL	6			
Symphoricarpos orbiculatus, coralberry	2–4	P, S		FACU	1			
Symphyotrichum × amethystinum, amethyst aster (N.E. × heath)	2–3	P, S		FAC	5	11/6	10/8	
Symphyotrichum drummondii, Drummond's aster (2 var.)	2–4	P, S	A	FACU	3			
Symphyotrichum dumosum, rice-button aster (5 var.)	2–4	P, S, N	C30	FAC+	5		9/19	
**Symphyotrichum ericoides*, heath aster (3 var.)	3–5	P	A:D/M:D	FACU–	4	11/11	10/3	
Symphytotrichum racemosum, small white aster	2–4	P		FACW–	3			
**Symphyotrichum laeve*, smooth aster (4 var.)	2–4	P, S	A/M	UPL	8	9/26–11/5	9/19	
Symphytotrichum lanceolatum, panicled aster (2 ssp.)	2	P, S		OBL	3			
**Symphyotrichum lateriflorum*, side-flowered aster (6 var.)	2–4	P, S, W	A	FACW–	2		8/20	
**Symphyotrichum novae-angliae*, New England aster	2–3	P, S	A/M	FACW	4	11/9	10/4	DP*
**Symphyotrichum oblongifolium*, aromatic aster	4–5	P, N	O/O:C90	UPL	7	11/8–12/5	9/21	
**Symphyotrichum oolentangiense*, azure aster (2 var.)	3–5	P, S	A/M	UPL	7	10/28–11/22	10/3	
Symphyotrichum parviceps, small-headed aster	4–5	P, S		UPL	3			
Symphyotrichum patens, late purple aster (3 var.)	4–5	P, S		UPL	6			
**Symphyotrichum pilosum*, frost aster (2 var.)	2–5	P, S	C:D	FACU–	0	11/21–12/2	9/27	
Symphyotrichum praealtum, willow-leaved aster (4 var.)	2–3	P, S		OBL	4		9/15	
Symphyotrichum prenanthoides, crooked-stem aster	2–3	P, S, W	C:D	OBL	10		10/10	
Symphyotrichum puniceum, swamp aster (2 var.)	1–2	P, S	C	OBL	7		9/15	
Symphyotrichum racemosum, small white aster	2–4	P		FACW–	3			
Symphyotrichum sericeum, silky aster	4–5	P	C:G/M	UPL	9	10/13–11/15	9/20	
Symphyotrichum simplex, panicled aster (2 var.)	2–3	P, S	C	OBL	3		8/30	

(continued)

(continued)

	H_2O	Eco	Seed	Wet	STAT/CC	Seed-time	FLR-time	PREDTN
Symphyotrichum tubinellum, prairie aster	3–5	P, S	C/A	UPL	7			
Taenidia integerrima, yellow pimpernel	3–4	P, S, W, N	C/M	UPL	7	7/1–	5/16	
Talinum see *Phemeranthes*								
Tephrosia virginiana, goat's rue	5	P, S, N	H	UPL	7	9/1–9/20	6/29	
Tetraneuris herbacea, lakeside daisy	4–5	P, N		UPL	10			
**Teucrium canadense*, American germander (3 var.)	2–3	P, S	C/M	FACW–	3	9/7–10/15	6/10	
**Thalictrum dasycarpum*, purple meadow rue	2–3	P, S	C:G/M	FACW–	5	7/31–9/14	6/18	DP*
Thalictrum revolutum, waxy meadow rue	1–3	P, S		FAC	5		6/20	
Thaspium barbinode, hairy meadow parsnip	3–5	S, W		UPL	7			
Thaspium trifoliatum, meadow parsnip (2 var.)	3–4	P, S, W	M/C120:G	UPL	6		6/8	
Tradescantia bracteata, prairie spiderwort	4–5	P, N	M/C120:G	FACU–	E/7	6/11–7/1	5/16	
**Tradescantia ohiensis*, Ohio spiderwort	2–5	P, S	M/C120:G	FACU+	3	7/6–7/31	5/25	
Tradescantia virginiana, spiderwort	2–5	P, S	M	UPL	7		5/27	
Trautvetteria caroliniensis, false bugbane	2–4	P, S		FAC–	10			
Triadenum virginicum, marsh St. John's-wort	1–2	P		OBL	E/10			
Triantha glutinosa, false asphodel	1–2	P		OBL	10			
Trichostema brachiatum, false pennyroyal	4–5	P, S, N		UPL	7		8/20	
**Tridens flavus*, purple top (2 var.)	3–5	P, S, N	C	UPL	1	10/10–10/25		
Tridens strictus, longspike tridens	3–4	P		FACU	4			
Trifolium reflexum, buffalo clover (annual/biennial)	3–4	P, S, N		UPL	E/9		6/27	
Triglochin maritima, common bog arrow grass	1–2	P		OBL	10		6/25	
Triglochin palustris, slender bog arrow grass	1–2	P		OBL	10		7/10	
Triodanis perfoliata, venus' looking-glass	3–4	P, S, N	C	FAC	2		6/20	
Triplasis purpurea, purple sandgrass	4–5	P, N		UPL	6			
Tripsacum dactyloides, eastern gama grass	2–3	P	C/M	FAC+	4	8/8–10/3	7/7	
Valeriana edulis, common valerian (2 var.)	2	P, S	L/M	OBL	10		5/15	
Valerianella radiata, corn salad	2–4	P, S		FAC+	1			
Valerianella umbellicata, corn salad	2–3	P, S		FACW	5			

Veratrum virginicum, bunch flower	2–3	P, S	C	FACW+	10			
**Verbena hastata*, blue vervain (2 var.)	2–3	P, S	C:D/M:D	FACW+	3	9/12–10/25	7/4	
Verbena simplex, narrow-leaved vervain	3–5	P, N		UPL	4		7/25	
**Verbena stricta*, hoary vervain	4–5	P, S	C:D	UPL	2	10/20–11/10	8/12	
**Verbesina helianthoides*, sunflower crownbeard	3–4	P, S	C	UPL	6			
Verbesina virginica, frostweed (2 var.)	3–4	P, S		FACU–	6			
Vernonia arkansana, southern ironweed	2–3	P, S		FAC	10			
Vernonia baldwinii, Baldwin's ironweed (2 ssp.)	4–5	P, S		UPL	5			
**Vernonia fasciculata*, common ironweed (2 ssp.)	2–3	P, S	C	FACW	5	9/16–10/16	7/16	
Vernonia gigantea, tall ironweed (2 ssp.)	2–4	P, S		FAC	4		8/10	
Vernonia missurica, Missouri ironweed	2–4	P	C	FAC+	5		8/18	
**Veronicastrum virginicum*, Culver's root	2–4	P, S	A:D/M:D	FAC	6	9/25–10/22	7/5	DP
Vicia americana, American vetch (2 ssp. in US)	3–4	P, S, N	H/M	UPL	6		6/1	
**Viola* sp. (*cucullata*, *sororia*, and *affinis*), common blue violet	2–4	P, S, W	P:D	FAC	1	5/30–6/15	5/8	
Viola lanceolata, lance-leaved violet (3 ssp.)	2–3	P, N		OBL	7		5/20	
Viola × palmata, early blue violet	4	P, S	P:D	UPL	–			
Viola pedata, birdfoot violet	4–5	P, S, N	C:D/M:D	UPL	7		5/22	
**Viola pedatifida*, prairie violet	3–4	P	M:D	FACU–	9	6 & 8/14–10/23	5/7	
Viola × primulifolia, primrose violet	2–3	P, S		FACW+	10			
**Viola sagittata*, arrow-leaved violet (2 var.)	2–5	P, S, N	P:Q:D	FACW–	6		5/10	
Viola × viarum, plains violet	3–4	P		FACU	5		5/16	
Vulpia octoflora, six-weeks fescue (3 var.)	4–5	P		FACW–	2			
Zigadenus elegans, white camass (2 ssp.)	2	P, S		–	–		7/13	
Zigadenus venenosus v. *gramineus*, death-camass	3–4	P, S		UPL	10			
Zizania aquatica v. *aquatica*, wild rice	1	P		OBL	9			
Zizia aptera, heart-leaved golden alexanders	3–4	P, S	A:D	FACU	9	8/8–9/5	5/16	

(continued)

(continued)

	H_2O	Eco	Seed	Wet	STAT/CC	Seed-time	FLR-time	PREDTN
**Zizia aurea*, golden alexanders	2–4	P, S	O:G/M	FAC+	6	7/23–8/18	5/13	DP
Other peripheral plants occasionally found in prairies [woody species (trees), herbs mainly in savannah/woodlands, certain wetland species, plants in disturbed areas (weeds), and species found locally on the edges of the range]. Number of varieties or forms are not necessarily listed								
Acalypha gracilens, slender 3-seeded mercury	3–4	P, S		UPL	4		9/1	
Acalypha rhomboidea, three-seeded mercury	3–4	P, S		FACU	0			
Acalypha virginica, 3-seeded mercury	3–4	P, S		FACU	2		9/20	
Achillea millefolium v. *occidentalis*, wooly yarrow	2–4	P, S		FACU	0			
Acorus calamus, sweet flag	1	P		OBL	4			
Agrostis perennans, upland bent-grass	3–4	S		FAC–	4		8/4	
Alopecurus carolinianus, annual foxtail	2–3	P		FACW	0		5/21	
Ambrosia artemisiifolia, ragweed (annual, 3 var.)	3–5	P, S		FACU	0		9/3	
Ambrosia bidentata, lanceleaf ragweed (annual)	3–5	P		FACU–	0			
Ambrosia psilostachya, western ragweed (2 var.)	3–5	P, S		FAC–	2			
Ambrosia trifida, giant ragweed (annual, 2 var.)	2–3	P		FAC+	0		9/10	
Anemone quinquefolia, wood anemone (3 var.)	2–4	S, W		FAC	7		4/15	
Angelica venenosa, wood angelica	3–4	P, S, W		UPL	8			
Apios americana, ground nut	2–3	P, S.W		FACW	3			
Argemone albiflora, bluestem prickly poppy (2 ssp.)	4–5	P		UPL	*			
Argentina anserina, silverweed	2–3	P, S		FACW+	6		6/15	
Aristida oligantha, three-awn	3–5	P, S		UPL	0			
Arundinaria gigantea, giant cane (2 ssp.)	2	River banks		FACW	5			
Asclepias perennis, thin-leaved milkweed	1–2	P, S, W		OBL	10			
Asclepias syriaca, common milkweed	3–5	P, S	C	UPL	0		6/25	
Aureolaria pedicularia, clammy false foxglove	4–5	P, S, W		UPL	9			
Bartonia virginica, yellow bartonia	2	P, S		FACW+	10			
Bergia texana, Texas bergia	1–2	P, S		OBL	10			
Bidens frondosa, beggar-ticks	2–3	P, S		FACW	1		9/10	
Bidens vulgata, tall beggar-ticks	2–3	P, S		FACW	0		9/5	

Bignonia capreolata, tendrilled trumpet-flower	2–3	P, S, W		FACW	8	
Bromus pubescens, hairy wood chess	2–3	S		FACW–	7	7/25
Buchloe dactyloides, buffalo grass (from west)	4–5	P	A/M	FACU–	*	
Callirhoe digitata, fringed poppy-mallow	4–5	P		UPL	*	
Callirhoe involucrata, purple poppy-mallow (2 var.)	4–5	P		UPL	*	
Calystegia sepium, American bindweed (7 ssp.)	2–4	P, S		FAC	1	6/25
Campanulastrum americanum, tall bellflower	2–4	S, W		FAC	4	7/18
Carex crinita, fringed sedge (3 var.)	2–3	P, S, W	C:D	FACW+	8	6/28
Carex leavenworthii, Leavenworth's sedge	4–5	P, S, N		UPL	2	5/10
Carex lupuliformis, sedge	1–2	P, S, W		FACW+	5	7/15
Carex lupulina, thick sedge	1–2	P, S, W	C:D	OBL	5	6/10
Carex muskingumensis, muskingum sedge	1	S		OBL	6	6/1
Carex normalis, larger straw sedge	1–3	P, S, W	C:D	FACW	4	5/29
Carex pensylvanica, Penn sedge	3–4	S, W	C:D	UPL	5	5/5
Carex sparganioides, bur-reed sedge	3–4	S, W		FAC	4	
Carex sprengelii, long-beaked sedge	2–4	P, S, W	C:D	FAC	8	5/14
Carex swanii, downy green sedge	2–4	P, S, N		FACU	8	5/30
Carex tenera, straw sedge	1–3	P, S, W	C:D	FAC+	5	6/1
Carex typhina, cat-tail sedge	1–2	P, S, W	C	OBL	6	7/26
Carya sp., hickory (at least 2 sp.)						
Cenchrus longispinus, sand bur	4–5	P, N		UPL	0	
Cephalanthus occidentalis, buttonbush	1–2	P, S		OBL	4	
Chaerophyllum tainturieri, southern chervil	2–3	P, S		FACU+	1	
Chamaesyce glyptosperma, ridge-seeded spurge (annual)	4–5	P, N		UPL	3	9/15
Chamaesyce humistrata, hairy spreading spurge (annual)	2–3	P, S, N		FACW	1	
Chamaesyce maculata, blotched spurge (annual)	3–4	P, S		FACU–	0	8/20
Chamaesyce serpens, round-leaved spreading spurge (annual)	3–4	P, S		UPL	2	

(continued)

(continued)

	H_2O	Eco	Seed	Wet	STAT/CC	Seed-time	FLR-time	PREDTN
Cinna arundinacea, wood/sweet reed-grass	1–2	S, W		FACW	5			
Cirsium altissimum, tall thistle (biennial)	2–3	P, S		UPL	3			
Claytonia virginica, spring beauty	2–3	S, W		FACU	1			
Clematis pitcheri, Pitcher's leather flower	2–4	S		FACU	4			
Clematis virginiana, virgin's bower	2–4	S		FAC	3			
Comarum palustre, marsh cinquefoil	1–2	P, S		OBL	10		6/22	
Commelina virginica, Virginia dayflower	2–3	P, S, W		FACW	5			
Comptonia peregrina, sweet fern	3–5	P, S, N		UPL	9		4/30	
Conoclinium coelestinum, blue boneset	2–3	P, S, W		FAC+	3		9/18	
Conyza canadensis, horseweed	2–4	P, S		FAC–	0			
Conyza ramosissima, dwarf fleabane	3–4	P, S		UPL	1			
Corispermum americanum v. *americanum*, common bugseed	3–4	P, S, N		FACU	6		9/1	
Corispermum nitidum, small bugseed	4–5	P, S, N		UPL	6			
Cornus sp., dogwood (several sp.) *Corylus americana*, hazel-nut	2–4	P, S		FAC	4			
Crataegus sp., hawthorn (about 15 sp.)								
Croton capitatus, capitate croton (2 var.)	4–5	P, N		UPL	0		7/22	
Crotonopsis michauxii, crotonopsis	4–5	P, N		UPL	8			
Cuscuta cephalanthi, buttonbush dodder	3–4	P, S		UPL	5		9/1	
Cuscuta compacta, compact dodder (2 var.)	2–4	P, S, W		UPL	10			
Cuscuta coryli, hazel dodder	2–4	P, S		UPL	5			
Cyperus bipartitus, brook flat sedge	2	P, S		FACW+	4			
Cyperus diandrus, umbrella flat sedge	2	P, S		FACW+	7			
Cyperus esculentus, chufa/ field nut sedge	1–3	P		FACW	0			
Cyperus strigosus, straw-colored cyperus	1–2	P		FACW	0			
Cypripedium parviflorum pubescens, large yellow lady's slipper	2–4	S, W		FAC+	8			

Dalea leporina, foxtail dalea (annual)	2–3	P	UPL	*	
Dasistoma macrophylla, mullein foxglove	2–3	S	FACU–	7	
Delphinium tricorne, larkspur	3–4	S, W	UPL	6	5/14
Dichanthelium clandestinum, deer tongue grass	2–3	S, W	FACW	4	
Dichanthelium commutatum, panic grass	2–4	P, S, N	FAC	7	
Dichanthelium dichotomum v. *dichotomum*, forked panic grass	3–4	P, S, W	FAC–	6	
Dichanthelium meridionale, mat panic grass	4–5	P, S, N	UPL	7	
Dichanthelium sphaerocarpon, round-fruited panic grass (2 var.)	3–5	P, S	FACU	7	
Echinochloa muricata, barnyard grass (2 var.)	1–2	P	OBL	0	
Eclipta prostrata, yerba-de-tajo (annual)	2–3	P, S	FACW	2	8/15
Elymus riparius, river-bank wild rye	2–3	S	FACW	6	
Epilobium strictum, downy willow-herb	1–2	P, S	OBL	T/10	8/7
Equisetum sp., horsetail/scouring rush (about 7 sp.)					
Eragrostis capillaries, lace grass	2–3	P, S	FACW	5	
Eragrostis pectinacea, love grass (3 var.)	2–4	P	FAC	0	
Erechtites hieracifolia, fireweed	2–4	P, S	FACU	2	
Erianthus alopecuroides, silver plumegrass	2–3	P, S, N	FACU–	4	
Erianthus brevibarbis, brown plume grass (2 var.)	1–2	P, S	OBL	10	
Erigeron annuus, annual fleabane (annual)	2–3	P	FAC–	1	5/20
Eupatorium serotinum, late-flowering thoroughwort	2–3	P, S	FAC+	1	9/15
Euphorbia cyathophora, wild poinsettia	2–3	P, S	UPL	0	
Euphorbia dentata v. *dentate*, wild poinsettia	3–4	P	UPL	0	7/19
Euphorbia spathulata, blunt-leaved spurge (annual)	3–4	P, S	FACU	5	
Eurybia furcata, forking aster	4–5	S	UPL	9	8/29
Frasera caroliniensis, American columbo	3–4	P, S, W	UPL	8	
Froelichia floridana v. *campestris*, cottonweed	4–5	P, N	UPL	5	

(continued)

(continued)

	H_2O	Eco	Seed	Wet	STAT/CC	Seed-time	FLR-time	PREDTN
Galium concinnum, shining bedstraw	3–4	S, W		FACU	4			
Galium pilosum, hairy bedstraw	4–5	S, W		UPL	7			
Glyceria borealis, northern manna grass (n. IL)	1–2	P, S		OBL	E/10		6/15	
Glyceria canadensis, rattlesnake manna grass (n. IL)	1–2	P, S		OBL	10		7/20	
Glyceria grandis v. *grandis*, American manna grass	1–2	P, S		OBL	8		7/5	
Gratiola virginiana v. *virginiana*, round-fruited hedge-hyssop	1–2	P, S		OBL	5			
Grindelia squarrosa, broad-leaved gum-plant (3 var.)	3–4	P		FACU	*			
Hedeoma hispida, rough pennyroyal (annual)	4–5	P, S, N		UPL	2		6/15	
Hedeoma pulegioides, American pennyroyal (annual)	4–5	P, S, N		UPL	4		8/31	
Helianthus annuus, common sunflower (annual, west)	2–4	P		FAC–	*		9/1	
Heuchera americana, alum root (3 to 4 var.)	3–4	S, W		FACU–	7			
Hexalectris spicata, crested coral root	3–5	S, W		UPL	10			
Hieracium canadense, Canada hawkweed	4–5	P, S, N		UPL	5		8/11	
Hudsonia tomentosa, woolly beach-heather (2 var.)	4–5	P, N		UPL	E/9		6/30	
Hymenopappus artemisifolius, white-bracted hymenopappus	4–5	P, S, W, N		UPL	9		6/20	
Iodanthus pinnatifidus, purple/false rocket	2	S		FACW	6		6/20	
Ipomoea pandurata, wild sweet potato/man-of-the-earth	3–4	P, S		FACU	2		7/20	
Isoetes butleri, Butler's quillwort	2–3	P, S, N		OBL	10			
Isoetes engelmannii, Engelmann's quillwort	1–2	P, S		OBL	9			
Isoetes melanopoda, black-based quillwort	1–2	P, S		OBL	10			
Iva annua, sumpweed (3 var. annual)	2–3	P, S		FAC	0			
Juncus nodatus, stout rush	1–2	P		OBL	6			
Lactuca biennis, tall blue lettuce	2–3	S		FAC	4		8/23	
Lactuca canadensis, Canada lettuce	2–3	P, S		FACU+	1	9/1–	7/31	
Lactuca floridana, Florida lettuce (2 var.)	2–3	S		FAC–	4		8/20	
Lathyrus ochroleucus, pale vetchling	3–5	S, W		UPL	8			
Lechea minor, pinweed	4–5	P, S, N		UPL	8		9/16	

Leersia lenticularis, catchfly grass	1–2	P, S		OBL	5		
Leersia oryzoides, rice cutgrass	1–2	P		OBL	3		
Leersia virginica, white grass	2–3	S		FACW	4		
Lepidium virginicum v. *virginicum*, common peppergrass	3–4	P		FACU–	0		
Leptochloa fusca v. *fascicularis*, salt meadow grass	1–2	P		OBL	0		
Leptochloa panicea, red sprangletop (2 var.)	2	P, S, N		FACW+	5		
Lespedeza violacea, violet bush-clover	4–5	S, N		UPL	5		8/26
Lilium superbum, turk's cap lily	2–3	S, W		UPL	9	9/16–	
Lobelia inflata, Indian tobacco	2–4	P, S, W	C:D	FACU–	4		8/10
Lotus unifoliolatus v. *unifoliolatus*, prairie bird's-foot trefoil	4–5	P, N		UPL	*		
Ludwigia decurrens, upright primrose-willow (annual)	1–2	P, S		OBL	9		
Ludwigia glandulosa ssp. *grandulosa*, cylindric-fruited ludwigia	1–2	P, S		OBL	8		
Lycopus asper, western water horehound	1–2	P, S		OBL	*		
Malus sp., crabapple (about 2 sp.)							
Malvastrum hispidum, false mallow	4–5	P, N		UPL	5		
Mimulus alatus, sharp-winged monkey flower	1–2	P, S		OBL	6		
Mimulus glabratus v. *jamesii*, yellow monkey flower	1–2	P, S		OBL	E/9		6/25
Mimulus ringens v. *ringens*, monkey flower	1–2	P, S	C:D	OBL	5		7/24
Muhlenbergia asperfolia, scratchgrass	1–2	P, S, N		FACW	*		
Muhlenbergia mexicana, leafy satin grass	2–3	P, S		FACW	4		
Muhlenbergia racemosa, upland wild timothy	2–4	P		FACW	0		
Muhlenbergia schreberi, nimblewill	2–3	P, S, W		FAC	0		
Myosotis verna, early scorpion-grass	3–4	P, S, W		FAC–	3		
Nuttallanthus canadensis, blue toad-flax (annual/biennial)	4–5	P, S, N		UPL	4		5/22
Orobanche uniflora, one-flowered broom rape	3–4	S, W		UPL	8		
Oxalis dillenii, yellow wood sorrel	2–4	P, S, W		FACU	0		
Oxalis stricta, tall wood sorrel	3–5	P, S, W		FACU	0		6/10

(continued)

(continued)

	H_2O	Eco	Seed	Wet	STAT/CC	Seed-time	FLR-time	PREDTN
Packera glabella, cress-leaved grounsel	1–2	P, S		OBL	0		6/14	
Panicum capillare, witch grass	2–4	P		FAC	0			
Panicum dichotomiflorum, fall panicum (3 var.)	2–3	P		FACW–	0			
Paronychia fastigiata, low forked chickweed (3 var.)	3–5	P, S		UPL	5			
Pascopyrum smithii, wheatgrass	2–4	P		FACU–	*			
Paspalum dissectum, swamp bead grass (south)	1–2	P		OBL	E/8			
Paspalum floridanum, giant bead grass (south)	1–2	P, N		FACW	7			
Pedicularis lanceolata, swamp lousewort	1–2	P, S		FACW+	9		9/10	
Pediomelum argophyllum, silvery scurfy pea (west)	4–5	P		UPL	*		7/8	
Pediomelum esculentum, prairie turnip (west)	4–5	P	C10:H	UPL	*			
Penstemon calycosus, calico beard-tongue	2–4	P, S, W	C30:G	FACU	3		6/30	
Penthorum sedoides, Virginia stonecrop	1–2	P, S		OBL	2		7/21	
Phacelia gilioides, phacelia (annual)	3–5	P, S		UPL	E/9			
Phacelia purshii, Pursh's phacelia (annual)	2–3	P, S, W		FACU–	4			
Phlox divaricata, wild blue phlox	2–4	S	C	FACU	5		5/11	
Phlox paniculata, garden phlox	3–4	P, S		FACU	3		8/5	
Phyla lanceolata, fog-fruit (2 var.)	1–2	P, S, N		OBL	1		7/20	
Physalis longifolia v. *subglabrata*, smooth ground-cherry	3–4	P		UPL	0		7/15	
Piptatherum pungens, short-horned rice grass	4–5	P, S, N		UPL	10			
Plantago rugelii, red-stalked plantain (2 var.)	2–3	P, S		FAC	0		7/20	
Polanisia dodecandra, clammyweed (3 ssp.)	3–5	P, N		UPL	0		8/4	
Polygala cruciata v. *aquilonia*, cross milkwort (annual)	2	P, S, N		FACW+	9		8/1	
Polygonum buxiforme, boxwood/shore knotweed	2–4	P, S, N		UPL	0			
Polygonum erectum, erect knotweed (annual)	2–4	P, S		FACU	0		7/8	
Polygonum prolificum, leafy knotweed	2–4	P		FAC–	0			
Polygonum punctatum, smartweed (3 var.)	1–2	P		OBL	3		8/18	
Polygonum ramosissimum, bushy knotweed (2 var.)	2–3	P, N		FAC	0			
Polygonum setaceum, bog smartweed	1–2	P		OBL	7			

Populus deltoides, eastern cottonwood	2–3	P, S		FAC+	2	
Potentilla norvegica, rough cinquefoil	2–4	P, S		FAC	0	
Prenanthes alba, white lettuce	3–4	S, W		FACU	5	
Proboscidea louisianica, devilsclaw (west)	2–3	P		FAC+	*	
Prunella vulgaris lanceolata, self-heal	2–4	P, S		FAC	1	
Prunus sp., native plum (about 2 or more sp.)						
Ptilimnium nuttallii, laceflower	2–3	P		FACW+	E/7	
Quercus sp., oak (2 or more sp.)						
Ranunculus cymbalaria, seaside crowfoot	1–2	P		OBL	2	7/13
Ranunculus hispidus, bristly buttercup (3 var.)	1–2	P, S		FACW+	8	4/25
Rhus sp., sumac (several sp.)						
Rorippa sessiliflora, sessile-flowered cress	1–2	P, S		OBL	3	6/25
Rorippa sinuata, spreading yellow cress	2–3	P, S, N		FACW	3	
Rosa blanda, early wild rose (3 var.)	3–5	P, S, N	C:H:F	FACU	4	6/20
Rotala ramosior, tooth-cup	1–2	P, N		OBL	4	8/24
Rubus sp., blackberry (several sp.)						
Rudbeckia grandiflora v. *grandiflora*, large black-eyed susan	4–5	P		UPL	*	
Rudbeckia laciniata v. *laciniata*, green-headed coneflower	2–3	S	C/A	FACW+	3	8/29
Ruellia pedunculata, stalked ruellia (2 var.)	4–5	P, S, W		UPL	7	
Sagittaria latifolia, broad-leaved arrowhead	1	P		OBL	4	8/10
Salix sp., willow (about 17 species mainly in wetlands)						
Saururus cernuus, lizard's tail	1–2	P, S, W		OBL	5	7/24
Scrophularia lanceolata, hare figwort	3–4	S		FACU+	5	6/13
Silene nivea, snowy campion	2–3	S, W	C	FACW	8	7/15
Silphium trifoliatum, rosinweed (2 var.)	4–5	P, S		UPL	E/10	
Solanum carolinense v. *carolinense*, horse-nettle	2–4	P		FACU–	0	7/14
Solidago flexicaulis, zig-zag goldenrod	2–3	P, S, W	C	FACU	6	9/11

(continued)

(continued)

	H_2O	Eco	Seed	Wet	STAT/CC	Seed-time	FLR-time	PREDTN
Solidago gigantea, late goldenrod	2–3	P, S		FACW	3		8/28	
Solidago petiolaris, downy goldenrod (2 var.)	4–5	S, W		UPL	8			
Solidago radula, western rough golden-rod (3 var.)	4–5	S, N		UPL	7			
Solidago rugosa, rough goldenrod (2 ssp., 3 var.)	2–3	P, S		FAC+	8		9/15	
Solidago ulmifolia, elm-leaved golden-rod (3 var.)	3–4	P, S, W, N	C	UPL	5		9/1	
Spermacoce glabra, smooth button-weed	1–3	P, S, W		FACW+	4			
Spermolepis echinata, bristly-fruited spermolepis	4–5	P		UPL	0			
Spermolepis inermis, spermolepis/smooth scaleseed	4–5	P, S, N		UPL	4		6/13	
Spiraea tomentosa, steeple-bush	1–2	P, S	C:D	FACW	8		8/19	
Strophostyles helvola, trailing wild bean (annual)	2–3	S, N		FAC+	3		8/15	
Symphyotrichum anomalum, many-rayed aster	4–5	S, N		UPL	8			
Symphyotrichum boreale, rush aster	1–2	P		–	–		9/7	
Symphyotrichum ciliatum, rayless aster (west)	2–3	saline		FAC	*			
Symphyotrichum ontarionis, Ontario aster	2–3	P, S		FAC	4		10/1	
Thalictrum thalictroides, rue anemone	3–4	S	L	UPL	5		4/15	
Trichostema dichotomum, blue curls	4–5	P, S, N		UPL	6		9/14	
Triodanis leptocarpa, venus' looking-glass	3–5	P, S		UPL	8		6/20	
Triosteum angustifolium, narrow-leaved horse-gentian	3–4	S		UPL	7			
Triosteum perfoliatum, horse-gentian (2 var.)	2–4	P, S, W		UPL	5		6/10	
Typha sp., cat-tail (2 species)	1–2	P		OBL	1			
Verbena bracteata, creeping vervain	3–4	P, N		FACU	1		6/30	
Verbesina alternifolia, wingstem	2–3	P, S		FACW	4	/	9/1	
Veronica peregina, purslane speedwell	2–4	P		FACW+	0			
Veronica scutellata, marsh speedwell	1–2	P		OBL	9		6/23	
Viola nephrophylla, northern bog violet	1–2	P		OBL	8		5/5	
Xyris torta, slender yellow-eyed grass	1–2	P, N		OBL	10		7/20	

Glossary

Achene	A dry one-seeded indehiscent fruit.
Acute	Terminating in a sharp point.
Anther	The pollen-bearing part of the stamen.
Auricle	An ear-shaped lobe.
Basal	Bottom part of the plant.
Bract	A small leaf subtending a flower or flower cluster.
Capsule	Dry fruit of two or more carpels, usually dehiscent by valves or teeth.
Carpel	A pistil or one member of a compound pistil.
Chaff	Thin dry scales.
Coma	Tuft of hairs at ends of certain seeds.
Cordate	Heart-shaped.
Coriaceous	Leathery in texture.
Corm	A solid bulb.
Culm	The stem of grass or sedge.
Cuneate	Wedge-shaped.
Cylindrical	Like a cylinder.
Dehiscent	Opening to release the contents.
Dentate	Toothed, usually with outward directed teeth.
Downy	With fine soft hairs.
Elliptical	Form of an ellipse rounded on both ends.
Eurasia (n)	Originating from the continents of Europe and Asia.
Ellipsoids	An elliptical solid body.
Floret	Small flower, usually one of a cluster.
Follicle	A fruit composed of one carpel dehiscent along one suture, as in Apocynum.
Forbs	Flowering plants exclusive of grasses.
Glabrous	Without hairs, smooth.
Glaucous	Covered with a whitish bloom.
Globose	Spherical or nearly so.

Glyphosate	A broad-spectrum herbicide composed of the isopropylamine salt of N-(phosphonomethyl)-glycine.
Glume	Scaly bract of spiklets of grasses and sedges.
Glutinous	Sticky exudation.
Hirsute	Pubescent with coarse or stiff hairs.
Hoary	Grayish-white fine pubescence.
Incised	Cut sharply into lobes.
Inflorescence	The flowering part of plants.
Infrastipular	An appendage below the base of a petiole where it attaches to the stem.
Internode	Portion of a stem between two successive nodes.
Involucral	Belonging to a circle or collection of bracts surrounding a flower cluster or a single flower.
Lance/Lanceolate	Shaped like a lance, many times longer than wide, broadest toward the base and narrowed to the apex.
Mesic	Applied here as indicating medium or average soil moisture.
Metric, m, cm, mm	Metric is an international unit of measure in which 1 meter (m) equals 3.28 ft; 1 centimeter (cm) is 1/100 of a meter=0.39 inches; 1 millimeter (mm) is 1/1,000 of a meter=0.039 inches.
Mycorrhizal	Pertaining to soil fungi that associate with the roots of plants for their mutual benefit.
Node	The junction of a stem, usually hard and swollen, at which leaves are usually attached.
Nutlet	Diminutive nut.
Oblong	Two or three times as long as broad with parallel sides.
Oblique	Slanting or unequal-sided.
Obovate	Inverted ovate.
Oval	Broadly elliptical.
Ovate	Egg-shaped with the broad end down.
Palmate	Radiately lobed, divided leaf, like fingers.
Pappus	The bristles, awns, teeth, etc., surmounting the achene, such as in Compositae.
Peduncle	Stalk of a flower or flower cluster.
Perigynium/Perigynia	The inflated sac(s) that encloses the ovary in Carex sp.
Petiole	The supporting stalk of a leaf.
Phyllary	An involucral bract in the Compositae.
Pilose	With long soft hairs.
Pinnate	Compound leaf with the leaflets arranged on each side of a common axis.
Pistillate scale	A thin dry leaf associated with pistil.
Pith	The spongy center of a stem.

Pod A seedcase, particularly of legumes.
Protuberance Bulge, swelling.
Pubescent Covered with hairs, especially short, soft or downy hairs.

Rhizome A subterranean stem, usually rooting at the nodes, producing new plants from the rooted nodes.

Scabrous Rough.
Scarification Producing a hole in the seed coat.
Senesce The process of aging.
Sepal One of the leaves of a calyx (the outer green leaf of the flower surrounding the petals).
Serrated Having sharp teeth pointed forward.
Sessile Without a stalk of any kind.
Shatter The process of releasing seed.
Spiklet A secondary spike in grasses or sedges.
Stamen A pollen-bearing organ of the flower.
Stratification The process of mixing seed with moist sand to facilitate germination.
Stolon A runner or basal branch that will root.
Style The attenuated part of the pistil connecting the stigma and ovary.
Subtend To enclose in an angle as between a leaf and its stem.

Tomentose Densely pubescent with matted wool.
Turbinate Top-shaped, inversely conical.

Umbel An inflorescence in which the peduncles of a cluster emanate from the same level, such as in the Umbelliferae family, e.g., Zizia, Angelica sp.

References

1. McClain WE (1986) Illinois prairie: Past and future. A restoration guide. Publication of Illinois Department of Conservation, Springfield
2. Betz RF (1972) In: Korling T (ed) The prairie swell and swale. De Lange/van Leer NV, The Netherlands
3. Rock HW (1981) Prairie propagation handbook, 6th edn. Milwaukee County Department of Parks, Recreation and Culture, Milwaukee
4. Axelrod DI (1985) Rise of the grassland biome, central North America. Bot Rev 51:163–201
5. Cornelius DR (1947) The effect of source of little bluestem grass seed on growth, adaptation, and use in revegetation seedings. J Agric Res 74:133–143
6. Taft JB, Wilhelm GS, Ladd DM, Masters LA (1997) Floristic quality assessment for vegetation in Illinois: a method for assessing vegetation integrity. Erigenia 15:3–95
7. Halbleib B (2005) What might have been. Pa Game News 76(3):4–8
8. Panzer R, Gnaedinger K, Derkovitz G (2010) The prevalence and status of conservative prairie and sand savanna insects in the Chicago wilderness region. Nat Areas J 30:73–81
9. Tallamy DW (2007) Bringing nature home: How native plants sustain wildlife in our gardens. Timber Press, Portland
10. Packard S, Mutel CF (eds) (2005) The tallgrass restoration handbook for prairies, savannas, and woodlands, 2nd edn. Island Press, Washington
11. Prairie Moon Nursery 1993 to 2009 Catalog and Cultural Guide. www.prairiemoonnursery.com
12. Deno NC (1993) Seed germination theory and practice, 2nd edn; (1996) First supplement; (1998) Second supplement. Norman C. Deno, 139 Lenor Drive, State College, PA 16801
13. Britton NL, Brown A (1970) An illustrated flora of the Northern United States and Canada, vol 1–3, 2nd edn. Dover Publications, New York
14. Fernald ML (1987) Gray's manual of Botany, 8th edn. Dioscorides Press, Portland
15. Barkley TM, Brooks RE, Schofield EK (eds) (1986) Flora of the Great Plains. University Press of Kansas, Lawrence
16. Swink F, Wilhelm G (1974) Plants of the Chicago region, 2nd edn. The Morton Arboretum, Lisle
17. Mohlenbrock RH (1999) The illustrated flora of Illinois, Sedges: *Carex*. Southern Illinois University Press, Carbondale and Edwardsville
18. Buchanan BB, Gruissem W, Jones RL (eds) (2000) Biochemistry and molecular biology of plants. American Society of Plant Physiologists (Biologists), Rockville, pp. 797
19. Mohlenbrock RH (1973) The illustrated flora of Illinois: Grasses *Panicum* to *Danthonia*. Southern Illinois University Press, Carbondale and Edwardsville
20. Proksch A, Wagner H (1987) Structural analysis of a 4-*O*-methyl-glucuronoarabinoxylan with immuno-stimulating activity from *Echinacea purpurea*. Phytochemistry 26:1989–1993
21. Segall HJ, Wilson DW, Dallas JL, Haddon WF (1985) *Trans*-4-Hydroxy-2-hexenal: a reactive metabolite from macrocyclic pyrrolizidine alkaloid senecionine. Science 229:472–475

22. Mohlenbrock RH (1970) The illustrated flora of Illinois: Flowering plants, lilies to orchids. Southern Illinois University Press, Carbondale and Edwardsville
23. Weber E, Schmid B (1993) Abstract 2156, XV International Botanical Congress, Yokohama, Japan
24. Denison E (1978) Missouri wildflowers, 3rd edn. Missouri Department of Conservation, Jefferson City
25. Keener CS, Leitzel W (2008) Stemless blue violets. Notes of the Pennsylvania Native Plant Society 11(2):1, 6–9; see also Russell NH (1965) Sida 2(1):1–113
26. Curtis JT (1959) The vegetation of Wisconsin. University of Wisconsin Press, Madison
27. Pauly W, Coyle A (1977) Prairie seed collecting dates (May–September). Dane County Highway and Transportation Department, Madison
28. Thompson JR (1992) Prairies, forests, and wetlands: The restoration of natural landscape communities in Iowa. University of Iowa Press, Iowa City
29. Pauly WR (1985) How to manage small prairie fires, revised edn. Dane County Park Commission, Madison
30. Marshall P, Paskus L, Wagner R, Tallmadge A (2008) Forest magazine (Winter). FSEEE, Eugene, pp 14–24
31. Numerous authors (1990) Vegetation management manual: Overview of vegetation management guidelines. Illinois Nature Preserves Commission, Illinois Department of Conservation and Natural Resources, Springfield, IL
32. Vaughn SF, Berhow MA (1999) Allelochemicals isolated from tissues of the invasive weed garlic mustard (*Alliaria petiolata*). J Chem Ecol 25:2495–2504
33. Nuzzo VA (1991) Experimental control of garlic mustard [*Alliaria petiolata* (Bieb.) Cavara and Grande] in northern Illinois using fire, herbicide and cutting. Nat Areas J 11:158–167
34. Kurtz C (2001) A practical guide to prairie reconstruction. University of Iowa Press, Iowa City
35. Fromm K (2001) Principles and procedures of prairie establishment. Self-published. Available from Prairie Moon Nursery, Winona
36. Shirley S (1994) Restoring the tallgrass prairie: An illustrated manual for Iowa and the upper Midwest. University of Iowa Press, Iowa City
37. Mohlenbrock RH (1970) The illustrated flora of Illinois: Flowering plants, flowering rush to rushes. Southern Illinois University Press, Carbondale and Edwardsville
38. Mohlenbrock RH (1972) The illustrated flora of Illinois: Grasses, *Bromus* to *Paspalum*. Southern Illinois University Press, Carbondale and Edwardsville
39. Mohlenbrock RH (1976) The illustrated flora of Illinois: Sedges, *Cyperus* to *Scleria*. Southern Illinois University Press, Carbondale and Edwardsville
40. Mohlenbrock RH (1978) The illustrated flora of Illinois: Flowering plants, Hollies to Loasas. Southern Illinois University Press, Carbondale and Edwardsville
41. Mohlenbrock RH (1980) The illustrated flora of Illinois: Flowering plants, Willows to Mustards. Southern Illinois University Press, Carbondale and Edwardsville
42. Mohlenbrock RH (1981) The Illustrated Flora of Illinois: Flowering Plants, Magnolias to Pitcher plants. Southern Illinois University Press, Carbondale and Edwardsville
43. Mohlenbrock RH (1982) The illustrated flora of Illinois: Flowering plants, Basswoods to Spurges. Southern Illinois University Press, Carbondale and Edwardsville
44. Hitchcock AS (1971) Manual of the grasses of the United States, vol 1, 2, 2nd edn. Dover Publications, New York
45. Herkert JR (ed) (1991) Endangered and threatened species of Illinois: Status and distribution, vol 1: Plants. Illinois Endangered Species Protection Board, Springfield

Native Index

Alien Index

Scientific Native Index

Alien Species Index

MIX
Papier aus verantwortungsvollen Quellen
Paper from responsible sources
FSC® C105338

FSC
www.fsc.org

If you have any concerns about our products,
you can contact us on
ProductSafety@springernature.com

In case Publisher is established outside the EU,
the EU authorized representative is:
Springer Nature Customer Service Center GmbH
Europaplatz 3, 69115 Heidelberg, Germany

Printed by Libri Plureos GmbH
in Hamburg, Germany